RECENT TECHNOLOGIES IN CARBON SEQUESTRATION AND CLIMATE CHANGE RISK MANAGEMENT

Editors

P. Christy Nirmala Mary

R. Shanmugasundaram

R. Muruga Ragavan

RECENT TECHNOLOGIES IN CARBON SEQUESTRATION AND CLIMATE CHANGE RISK MANAGEMENT

First Edition: November 2020
Printed: January 2021

Published by: Tamil Nadu Agricultural University, Coimbatore

Editors: P. Christy Nirmala Mary, R. Shanmugasundaram & R. Muruga Ragavan

ISBN: 978-93-90082-74-2

Publisher

SHANLAXPUBLICATIONS
61, 66 T.P.K. Main Road
Vasantha Nagar
Madurai – 625003
Tamil Nadu, India

Ph: 0452-4208765,
Mobile: 7639303383
email:publisher@shanlaxpublications.com
web: www.shanlaxpublications.com

PREFACE

Carbon sequestration is a long term removal, capture or sequestration of carbon dioxide from the atmosphere to slow or reverse atmospheric CO_2 pollution and to mitigate or reverse *global warming*. In response to growing concerns about climate change resulting from increased carbon dioxide concentrations in the atmosphere, considerable interest has been drawn to the possibility of increasing the rate of carbon sequestration through changes in land use and forestry and also through geoengineering techniques such as carbon capture and storage.

According to the Intergovernmental Panel on Climate Change (IPCC), improved agricultural practices and forest-related mitigation activities can make a significant contribution to the removal of carbon dioxide from the atmosphere at relatively low cost. These activities could include improved crop and grazing land management—for instance, more efficient fertilizer use to prevent the leaching of unused nitrates, tillage practices that minimize soil erosion, the restoration of organic soils, and the restoration of degraded lands.

The need of hour focuses on understanding the effects of climate change on livelihood realities and development perspectives of especially vulnerable population groups. Impact and vulnerability assessments provide an important basis for the identification of adaptation requirements as well as analyses of loss and damage.

A compilation about the carbon sequestration methods and climate change risk assessment would be a guide for understanding the present scenario of the world environment. Hence the chapters presented in this book envisages the novelties about conserving ecosystem by sequestering carbon in the prevailing ecosystem for mitigation of climatic fluctuations and to maintain the reduced emission strategies to protect the ecosystem from adverse climatic variables. This book is a treasure for under and post graduates and research scholars.

Authors

TAMIL NADU AGRICULTURAL UNIVERSITY

Prof. N. KUMAR, Ph.D., F.H.S.I., F.S.P.H.,
Vice-Chancellor

Coimbatore - 641003
Tamil Nadu, India

FOREWORD

India is vulnerable to climate change, notably due to the melting of the Himalayan glaciers and changes in the quantity, spread and reliability of monsoon. India is the world's third largest emitter of greenhouse gases (GHGs), after China and the US. India's total emissions are 7% of global emissions and are increasing at 4.5% per annum.

India's current and expected future emissions are sufficiently massive and may likely to have an adverse effect which need to be addressed. Integrating tree cultivation in multiple land uses will support achieving India's Nationally Determined Contribution (NDC) to sequester additional 2.5 to 3 billion tonnes carbon dioxide (CO_2) equivalent by 2030 as per the Paris Climate agreement.

The book on "Recent Technologies for Carbon Sequestration and Climate Change Risk Management" contains chapters which provide basic information about the recent technologies for carbon sequestration in terrestrial, forest, grassland, agro-forestry, agricultural ecosystem in addition to the climate change mitigation scenario of the world. The vegetation carbon sink sources were identified and mapped by applying remote sensing tools and technologies are deployed to know how the sequestering potential of carbon in natural and man engineered ecosystem varies for mitigating climate change.

This book will give a wider knowledge about the climate change mitigation strategies mainly through carbon sequestration and capture.

I congratulate the authors who have included the novelties about climate change mitigation strategies for conserving natural environment.

(N.KUMAR)

Date: 18.01.2021
Place: Coimbatore

● Tel : Off.: +91 422 2431788   ●Res. : +91 422 2430887   ●Fax: +91 422 2431672   ●Email: vctnau@tnau.ac.in

CONTENTS

Chapter 1

BIOGEOCHEMICAL SEQUESTRATION OF CARBON IN PLANTS

P. Christy Nirmala Mary

Associate Professor (SS&AC), Dept. of Soils and Environment, Agricultural College and Research Institute, TNAU, Madurai

INTRODUCTION

The global concentration of atmospheric CO_2 has increased significantly from ~280 to 391 ppmv since 1750 as a result of human activities such as fossil fuel combustion, deforestation, biomass burning and land use change (Macías and Arbestain, 2010). As global increase of atmospheric CO_2 concentration may cause dangerous climate change (IPCC, 2007). Various approaches that can securely reduce and sequestrate carbon emissions are being pursued,and among the most promising is the terrestrial biogeochemical carbon sequestration (IPCC, 2007). As an important long-term terrestrial carbon sequestration mechanism, biogeochemical sequestration of carbon within phytoliths may play a significant role in the global carbon cycle and climate change. One relatively stable form of organic carbon that is bio geochemically sequestrated within the silica bio mineralization features of terrestrial plants and that can accumulate in soil after the decomposition of that vegetation is the phytolith occluded carbon (PhytOC) fraction (Parr and Sullivan,2010). The PhytOC is more stable and can sustain much longer than other organic carbon fractions in the soil because of its strong ability to resist decomposition (Wilding, 1967). As an important part of terrestrial carbon, it can reach 82% of total carbon in some soil and sediments after 2000 years of litter leaf decomposition. Phytoliths, also called silica phytoliths, are non crystalline minerals that deposit inside cells and cell walls of different parts of plants when soluble silica is absorbed by the roots .Phytoliths have great potential to sequestrate atmospheric CO_2 through the formation of PhytOC (Parr *et al.*, 2009) and play a crucial function in the long-term terrestrial carbon cycle and climate change .The fluxes of millet, wheat and sugarcane for phytolith carbon bio-sequestration range up to 0.038 (Zuo and Lu, 2011), 0.246 and 0.36 $t\text{-}e\text{-}CO_2\ ha^{-1}\ a^{-1}$ respectively. In particular, the flux of carbon occluded within phytoliths of bamboo ranges up to 0.709 $t\text{-}e\text{-}CO_2\ ha^{-1}\ a^{-1}$, and current global bamboo forests (22 million ha) can securely sequestrate $1.56 \times 107\ t$ of atmospheric CO_2 per year. Researchers have suggested that if all potentially arable land (4.1 billion ha) is exploited to grow bamboo or other similar grass crops, the global potential of phytolith carbon bio-sequestration would approximately be 1.5 billion $t\text{-}e\text{-}CO_2\ a^{-1}$ equivalent to 11% of the current increase in atmospheric CO_2 (Parr *et al.*, 2010). The data indicated that the management of plants with high PhytOC content to maximize biomass production could adequately improve the secure terrestrial carbon sequestration. The widespread wetland ecosystem with fast plant growth and high biomass is an important terrestrial carbon sink, and plays an important role in global carbon cycle (Martha,2003) and global climate change . As one of the important constituents of terrestrial ecosystems, wetlands store 15%

of the total terrestrial carbon though it occupies only about 1% of the terrestrial surface (Zhang, 1999). Furthermore, the wetlands are mainly dominated by Poaceae, which are known to be proficient silica accumulators. However, to our best knowledge, the potential of wetland phytoliths in the long-term biogeochemical sequestration of atmospheric CO_2 has not been quantified globally and even regionally.

BIO GEOCHEMICAL SEQUESTRATION IN PADDY CROP

Rice is a typical silicon-accumulating plant. Silicon (Si), deposited as phytoliths during plant growth, has been shown to occlude organic carbon, which may prove to have significant effects on the biogeochemical sequestration of atmospheric CO_2. A promising biogeochemical means of reducing atmospheric CO_2 is the occlusion of carbon within plant phytoliths, also known as silica phytoliths or plant opals. Silicon is stored in plants mainly in the form of phytoliths. These silica phytoliths comprise non crystalline silica minerals deposited within cells and cell walls of different plant organs when monosilicic acid $[Si(OH)_4]$, is taken up by plant roots and transported to the aboveground organs, where phytoliths form near evaporative surfaces by deposition and polymerization. The silica phytoliths that are formed occlude some of the organic carbon that is extracted from atmospheric CO_2 during photosynthesis, which is then deposited during plant growth. This carbon fraction has been recognized as an important long-term terrestrial carbon (C) sink, sequestering about 1.5 billion tonnes of CO_2 annually. To evaluate the C occlusion within phytoliths during rice cropping as affected by silicate fertilizer additions, a study was taken with five silicate fertilizer additions in southern China in rice (Zimin *et al.*, 2014).A correlation also exists between the phytolith content in plants and the phytolith elemental content (Phytolith vs. Phytolith-Al, $R^2$0.6518; Phytolith vs. Phytolith-Fe, R^2 0.2973; Phytolith vs. Phytolith-Mg, $R^2$0.3194).

Fig .1. Representative scanning electron microscope (SEM) images of phytoliths from rice organs (A) grains, (B) leaf, (C) sheath, and (D) stem.

This can be explained that phytolith solubility of rice cultivars more controlled by difference in morphology (specific surface area). This documents that rice organs and cultivar can largely influence Si dynamics in soil-plant systems through variable solubility of phytoliths (Fig.1).

Table1.Experimental sites, Percentage of phytolith contents, percentage of carbon contents in phytolith and percentage of PhytOC content in straw.

Site	Treatment	Percentage of phytolith contents (%)	Percent carbon contents in phytoliths (%)	Percent PhytOC in straw (%)
LH	CK	4.30 ± 0.12d	4.03 ± 0.40a	0.17 ± 0.02b
	Si300	4.97 ± 0.33cd	3.33 ± 0.25ab	0.17 ± 0.02b
	Si600	5.60 ± 0.46bc	2.35 ± 0.19cd	0.13 ± 0.01b
	Si900	6.73 ± 0.47ab	2.11 ± 0.05d	0.14 ± 0.01b
	Si1500	7.73 ± 0.47a	2.14 ± 0.12d	0.17 ± 0.02b
	Si3000	7.60 ± 0.40a	3.06 ± 0.18bc	0.23 ± 0.02a

LST	CK	4.93 ± 0.13cd	3.09 ± 0.32ab	0.15 ± 0.02c
	Si300	5.47 ± 0.27c	3.65 ± 0.13a	0.20 ± 0.02b
	Si600	4.53 ± 0.35cd	2.63 ± 0.19bc	0.12 ± 0.02c
	Si900	5.60 ± 0.00c	2.15 ± 0.04c	0.12 ± 0.00c
	Si1500	6.73 ± 0.35b	2.18 ± 0.13c	0.15 ± 0.01c
	Si3000	8.20 ± 0.31a	3.43 ± 0.06a	0.28 ± 0.01a
XY	CK	6.53 ± 0.84b	3.91 ± 0.27bc	0.26 ± 0.05a
	Si300	6.27 ± 0.07b	4.59 ± 0.20a	0.29 ± 0.02a
	Si600	6.00 ± 0.12b	4.52 ± 0.14ab	0.27 ± 0.01a
	Si900	6.30 ± 0.26b	3.77 ± 0.27c	0.24 ± 0.02a
	Si1500	6.27 ± 0.27b	4.07 ± 0.09abc	0.26 ± 0.02a
	Si3000	7.80 ± 0.31a	3.97 ± 0.21abc	0.31 ± 0.02a
DZ	CK	6.40 ± 0.12b	2.11 ± 0.09e	0.14 ± 0.00d
	Si300	7.25 ± 0.05ab	2.56 ± 0.04d	0.19 ± 0.00c
	Si600	7.43 ± 0.20a	2.97 ± 0.05c	0.22 ± 0.01b
	Si900	7.33 ± 0.44ab	3.96 ± 0.09a	0.29 ± 0.01a
	Si1500	7.07 ± 0.26ab	3.41 ± 0.14bc	0.24 ± 0.00b
	Si3000	7.90 ± 0.59a	3.04 ± 0.23bc	0.24 ± 0.01b
QH	CK	6.93 ± 0.55b	4.46 ± 0.10a	0.31 ± 0.03b
	Si300	8.53 ± 0.58ab	4.47 ± 0.24a	0.38 ± .01ab
	Si600	8.33 ± 0.37ab	4.37 ± 0.16a	0.37 ± .03ab
	Si900	9.60 ± 1.20a	4.25 ± 0.54a	0.42 ± .10ab
	Si1500	10.13 ± 0.37a	4.38 ± 0.12a	0.44 ± 0.01a
	Si3000	10.40 ± 0.61a	4.41 ± 0.22a	0.46 ± 0.01a

Addition of silicate fertilizer affects PhytOC sequestration by improving PhytOC contents in rice straw and the plants growth; the average above-ground phytolith-occluded carbon (PhytOC) production rates at five of China's paddy rice production areas are increased by the addition of silicate fertilizer. Therefore, regulating Si supply during rice growth may serve as an effective tool in improving PhytOC production rate and play an important role in carbon bio-sequestration and global warming mitigation. The percent phytolith contents in rice straw at the five sites ranged from 4.30% to 10.40% (Table 1). The percentages of C contents in phytoliths ranged from 2.11% to 4.59% at the five sites tested and tended to decrease with increasing rice straw silica contents (Zimin ,2014).

BIOGEOCHEMICAL CARBON SEQUESTRATION WITHIN MILLET

Common millet and foxtail millet have become typical dry-farming crops in China since the Neolithic Age. The study of carbon conservation within phytoliths in these crops could provide insights into anthropogenic influences on the carbon cycle. The common millet and foxtail millet contained 0.136% ± 0.070% and 0.129% ± 0.085% phytolith-occluded carbon (PhytOC) on a dry mass basis, respectively (Falkowski *et al.*, 2000). Global Change based on the mean annual production of common millet and foxtail millet in the last 10 years, the phytolith occluded carbon accumulation rate of common millet and foxtail millet was approximately 0.023 ± 0.015 and 0.020 ± 0.010 t CO_2 ha^{-1} , respectively (Kosten *et al.*, 2010) assuming similar phytolith occluded carbon accumulation rate as for common millet (the highest accumulation rate was 0.038 t CO_2 ha $^{-1}$), this could result in the sequestration of 2.37 × 106 t CO_2 per year for the 62.4 ×106 ha^{-1} dry-farming crops in China. The total phytolith carbon sequestration rate was 7×106 t CO_2 within the 60-year period. However, phytolith occluded carbon has not yet been fully considered as a global carbon sink. Also, this carbon fraction is probably one of the best candidates for the missing carbon sink (Fig.2.) .

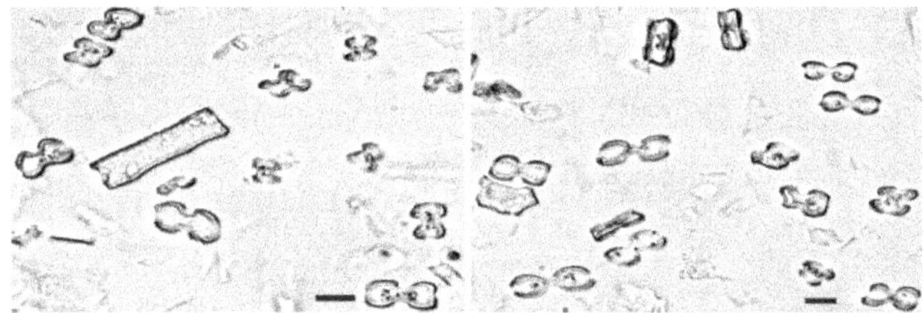

Fig.2. Phytoliths extracted from the millet samples (Both are from foxtail millet and black bar represents 10 μm).

The highest PhytOC sequestration in the last 60 years results (Figure 3) showed that $7×10^6$ t CO_2 are likely to have been sequestered and this would be sufficient to offset CO_2 emission released by the combustion of 2.69×10^6t standard coal. The PhytOC sequestration of foxtail millet generally declined during 1949 to 2008 in conjunction with changes in its planting area. However, other PhytOC crops such as wheat can capture more CO_2 from the atmosphere and securely sequester it for a long time.

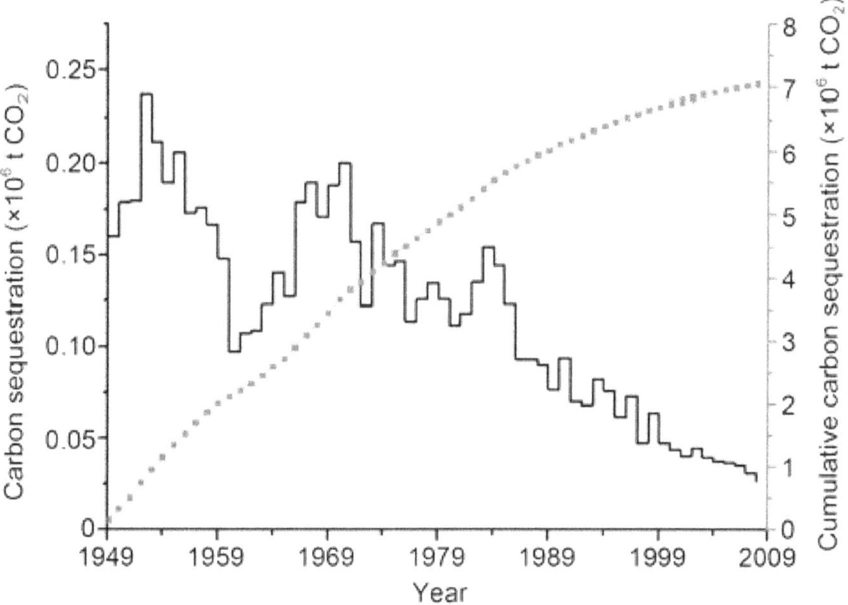

Figure 3. PhytOC accumulation rate of foxtail millet during 1949 to 2008.

SUGARCANE PHYTOLITH IN GEOCHEMICAL CARBON SEQUESTRATION

Carbon capture and long-term storage using silica phytoliths on newly planted and ratooned sugarcane varieties results indicated that there was significant variation in the phytolith occluded carbon (PhytOC) content of different varieties. The carbon content of the varieties tested under the particular suite of environmental conditions for which they were grown ranged from 0.12 te-CO_2ha-y^{-1} to 0.36 te-CO_2 ha-y^{-1}.

Table.2. Silica Phytolith, Phytolith , phytOC yield in sugarcane plant Variety new **(N) ratoon (R), Si-Phytolith content as a percentage of plant weight, carbon**

contents of phytoliths, PhytOC contents of the sugarcane varieties, PhytOC in carbon dioxide equivalents (e-CO$_2$) per hectare each year, and their value based on predicted opening prices for the Australian Emissions Trading Scheme 2010 in USD.

Sugarcane cultivar	Mass of dry plant material (g)	Mass remaining after digestion (g)	Si-Phytolith in plant material (%)	Carbon content of isolated phytoliths (%)	PhytOC yield (proportion of dry material) (%)	PhytOC yield* t/ e-CO$_2$	Price per ha at at $56.09 USD per tonne
N-1	8.0483	0.1288	1.6%	12.35	0.9940	0.2893	$16.23
N-2	8.0599	0.2134	2.6%	6.06	0.4884	0.2349	$13.17
R-3	7.7758	0.1495	1.9%	8.51	0.6615	0.2395	$13.43
R-4	8.0499	0.1740	2.2%	3.88	0.3126	0.1229	$ 6.89
N-5	8.0157	0.1965	2.5%	9.56	0.7663	0.3431	$19.24
R-6	8.0580	0.1497	1.9%	11.81	0.9516	0.3212	$18.02
R-7	8.0431	0.1177	1.5%	11.21	0.9016	0.2402	$13.47
N-8	7.7824	0.1581	2.0%	11.66	0.9074	0.3468	$19.45
R-9	8.0388	0.1038	1.3%	19.26	1.5483	0.3641	$20.42
N-10	7.7792	0.1721	2.2%	8.40	0.6535	0.2721	$15.26

* assumes typical dry biomass production of 40 tonnes ha^{-1} yr^{-1} for sugarcane on this property.

The lowest PhytOC yielding variety o f sugarcane was 0.12 t e-CO$_2$ ha-y^{-1} and the highest was 0.36 t e- CO$_2$ ha-y^{-1} While the PhytOC content of sugarcane varieties examined in this study (Table 2.) were lower than the 0.66 t e-CO$_2$ ha-y^{-1} reported for one variety in a previous study (Parr and Sullivan, 2005) there were significant quantities of carbon encapsulated in the cultivars examined. Thus if a farmer previously growing the lowest PhytOC yielding variety then chose to grow the highest PhytOC yielding variety in this study there would be a net increase of 0.24 T e-CO$_2$ ha-y^{-1} of carbon sequestered in phytoliths. Based on predicted opening prices ($56.09 USD per tonne) for the Australian Emissions Trading Scheme to come online in 2010, this additional carbon would be worth around $13.53USD per hectare to the farmer or cooperative (Parr *et al* ., 2009).

GEOCHEMICAL CARBON SEQUESTRATION IN BAMBOO

The total aboveground biomass (including leaves, branches, and culm) ranged from 20.82 to 48.68 t ha^{-1} per dry weight across the eight species, with the highest in PAKK [*Pleioblastus amarus* (Keng) Keng f] and lowest in PP (*Phyllostachys prominens* W. Y. Xiong) (Table 3.). The biomass of the rhizome was much smaller than that of the aboveground across the eight species with the exception that the biomass of rhizome of PP was almost three times higher than that in the aboveground.

The biomass of the belowground trunk ranged from 1.19 to 7.18 t ha^{-1} across the eight species. The proportion of below ground biomass to total biomass varied from 18.64% for PHO (*Phyllostachys heteroclada* Oliver) to 74.91% for PP, with a mean of 39.41%. There was a significant ($p < 0.05$) variation in the concentrations of Si, phytolith, C concentration in phytolith, and PhytOC in belowground biomass among the eight bamboo species (Table *3).*

The concentration of Si and phytolith in the rhizome ranged from 8.43 g kg^{-1} for PHMP [*Phyllostachys heterocycla (Carr.) Mitford'Pubescens'*] to 21.05 g kg^{-1} for PGM (*Phyllostachys glauca McClure*), and from 11.20 g kg^{-1} for PAKK to 35.44 g kg^{-1} for PGM, respectively.

The C concentration in phytolith in the rhizome ranged from 11.02 g kg^{-1} for BPM (*Bambusa piscatorum McClure*) to 80.42 g kg^{-1} for PHMP. There were no significant differences in the C concentration in phytolith in the rhizome among the other seven bamboo species except PHMP.

Table-3. Silica , phytolith , carbon concentration and phytOC in different Bamboo species.

Organ	Bamboo species	Si (g·kg^{-1})	Phytolith (g·kg^{-1})	C concentration in phytolith (g·kg^{-1})	PhytOC/dry biomass (g·kg^{-1})
Rhizome	PHMP	8.43 ± 3.53b	14.69 ± 3.12bc	80.42 ± 16.87a	0.83 ± 0.38a
	PPP	14.51 ± 2.47ab	34.93 ± 7.39a	32.27 ± 19.17b	0.83 ± 0.54a
	PP	10.58 ± 5.83b	24.53 ± 9.55ab	28.33 ± 2.52b	0.38 ± 0.06a
	PAMK	16.16 ± 9.17ab	19.78 ± 3.96bc	34.63 ± 27.20b	0.67 ± 0.43a
	PGM	21.05 ± 15.00a	35.44 ± 17.84a	23.20 ± 19.70b	0.66 + 0.50a
	PAKK	10.47 ± 3.93b	11.20 ± 5.33c	32.14 ± 2.75b	0.58 ± 0.08a
	PHO	10.70 ± 2.90ab	20.61 ± 5.57bc	23.94 ± 18.32b	0.53 ± 0.34a
	RMBPM	13.90 ± 4.69ab	34.74 ± 9.77a	11.02 ± 2.21b	0.34 + 0.04a
Belowground trunk	PHMP	4.88 ± 3.75bc	10.82 ± 0.78ab	57.25 ± 23.53bcd	0.31 ± 0.05bcd
	PPP	2.30 ± 0.60c	10.68 ± 1.58ab	179.99 ± 40.04a	0.94 ± 0.53a
	PP	3.87 ± 1.03bc	6.13 ± 0.66c	77.71 ± 67.83bcd	0.10 ± 0.02d
	PAMK	5.62 ± 2.25b	11.95 ± 4.57ab	38.51 ± 35.06cd	0.19 ± 0.08d
	PGM	5.12 ± 1.23bc	9.91 ± 5.45bc	138.58 ± 139.21ab	0.25 ± 0.04cd
	PAKK	6.89 ± 1.70b	9.10 ± 3.46bc	132.66 ± 82.84abc	0.54 ± 0.16bc
	PHO	14.07 ± 3.67a	14.95 ± 2.79a	40.73 ± 18.44cd	0.61 ± 0.13b
	RMBPM	5.05 ± 1.34bc	5.88 ± 1.19c	23.44 ± 5.57d	0.21 ± 0.06d

Values are means ± standard error of four replicates. Means followed by different letters within a column are significantly different at the p < 0.05 level. PHMP, Phyllostachys heterocycla (Carr.) Mitford 'Pubescens'; PPP, Phyllostachys praecox C. D. Chu 'Prevernais'; PP, Phyllostachys prominens W. Y. Xiong; PAMK, Pseudosasa amabilis (McClure) Keng f; PGM, Phyllostachys glauca McClure; PAKK, Pleioblastus amarus (Keng) Keng f; PHO, Phyllostachys heteroclada Oliver; BPM, Bambusa piscatorum McClure.

BIOGEOCHEMICAL SEQUESTRATION PHYTOLITHS IN GRASSLANDS

Grassland is one of the most widespread vegetation types worldwide, occupying more than one fifth of the world's land surface (Hall *et al.*, 1998). Grassland ecosystems may play an important role in the global terrestrial production of phytoliths due to their large area, high net primary productivity (NPP), and high Si concentration (Carnelli *et al.*, 2001; Blecker *et al.*, 2006).

As grassland ecosystems may contribute as much as 20% of the total terrestrial NPP (Scurlock& Hall, 1998) and phytolith content in grassland communities is as high as or higher than that in other ecosystems (e.g. woodlands), it can be inferred that they may contribute at least 20% of the total terrestrial phytolith production rate. However, large uncertainties still exist in estimates of grassland phytolith production, and its contribution to global terrestrial phytolith production (Carnelli *et al.*, 2001).

The average above-ground phytolith production rate for world grasslands is 752.3 10^6 t yr^{-1} or 7–21% of global terrestrial ecosystems estimated by Conley (2002). The average above-ground phytolith production rates of five grassland types of China (5.3 10^6t yr^{-1}or0.7% of world grassland rate) and China's grasslands in total (10.9 10^6 t yr^{-1} or 1.45% of world grassland rate) are much lower than those of North American non woody grasslands (17.4 10^6 t yr^{-1} or 2.32% of world grasslands). Silicon content and solubility of phytoliths (Fig.3.) ,the average silicon content of the spruce needles was 5.3 $\pm$ 0.5 g kg^{-1}. With an exponential increase above pH 11, dissolution curves of the silica gel and the diatomite were similar to the expected theoretical Si dissolution pattern (Zsuzsa Lisztes-Szab ,2004).

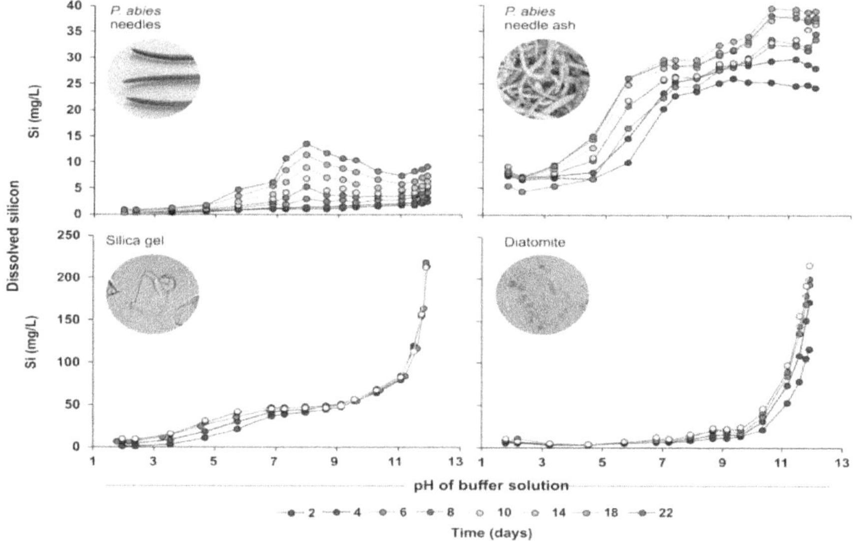

Fig. 3. Dissolved silicon content of the solution of spruce needles, ash of needles, silica gel and diatomite at pH 2–12.

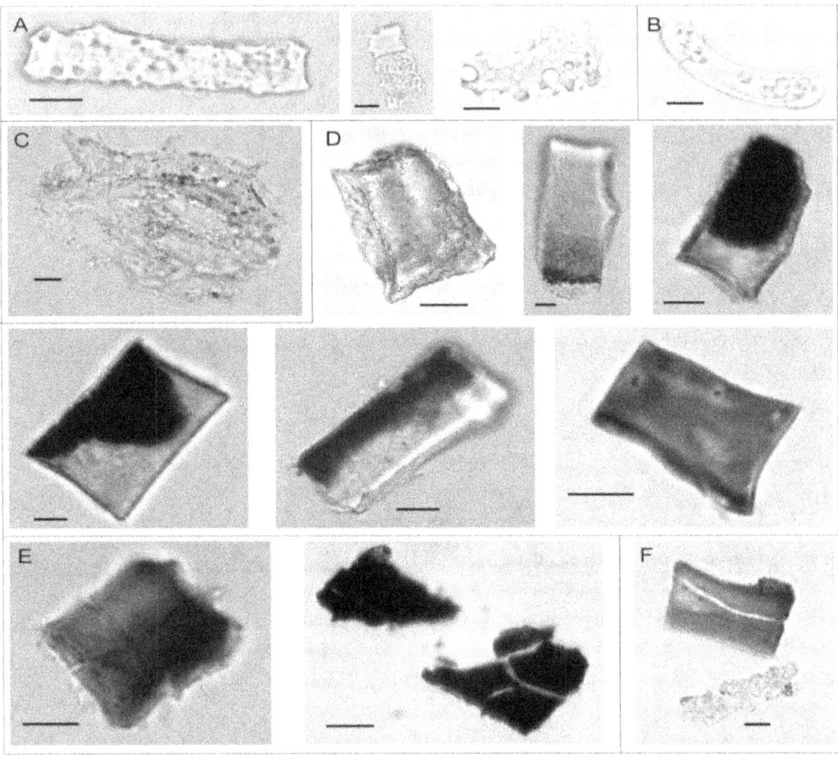

Fig. 4 Partially dissolved spruce phytoliths

Table .4. Comparison of estimated PhytOC fluxes and global total PhytOC rate in different ecosystems

CROP	PhytOC Content (dry weight) (%)	PhytOC sequestration fluxes (t-e-CO_2 ha^{-1} a^{-1})	sequestration rates (t-e-CO_2 a^{-1})
Fen	0.01–0.25	0.003–0.077	1.14×10^7
Rice	0.04–0.28	0.026–0.125	1.94×10^7
Bamboo	0.24–0.52	0.008–0.709	1.56×10^7
Sugarcane	0.31–1.54	0.12–0.36	0.72×10^7
Wheat	0.06–0.60	0.006–0.246	5.3×10^7
Millet	0.04–0.27	0.008–0.038	0.27×10^7

According to the published fen area (1.37×10^7 ha) by NBSC (2011) and our studied (Table 4) PhytOC sequestration flux (0.003–0.077 t-e-CO_2 ha^{-1} a^{-1}) from herb-dominated fen plants, it is estimated that the potential rates of CO_2 occluded within phytoliths of herb-dominated fen plants vary from 0.04×10^6 to 1.05×10^6 t CO_2 equivalents per year in China. Taking the world fen area (1.48×10^8 ha) and the largest phytolith carbon bio-sequestration flux (0.077 t-e-CO_2 ha^{-1} a^{-1}) of CO_2 occlusion within phytoliths from herb-dominated fen plants, about 1.14×10^7 t CO_2 equivalents per year would have been sequestrated in phytoliths of fen plants globally. As for the 5.7×10^8 ha of the world's wetlands , assuming a similar phytolith carbon bio-sequestration flux of 0.077 t-e-CO_2 ha^{-1} a^{-1}, the global potential rate for phytoliths carbon sequestration is estimated to be 4.39×10^7 t-e -CO_2 a^{-1}.

CONCLUSION

Agricultural activities have a complicated interaction with the global carbon balance. However, current studies on terrestrial ecosystem carbon sinks pay little attention to carbon sequestration by the PhytOC pool which has been poorly integrated into our current understanding of the carbon cycle. The potential of PhytOC sequestration provides a new approach for enhancing the soil organic carbon sink. More studies on the PhytOC of different crops are needed and this carbon fraction should now be incorporated in future long-term global carbon sequestration estimates. The soil amendment with plant residues and/or plant biochar should therefore be carried out taking into account the phytolith solubility of different plant organs of Si-high accumulator plants (e.g., leaves and sheaths of rice) and play an important role in carbon bio-sequestration and global warming mitigation.

References

- Carnelli AL, Madella M, Theurillat J-P (2001) Biogenic silica production in selected alpine plant species and plant communities. Annals of Botany, 87,425–434.

- Falkowski P, Scholes R J, Boyle E, et al. The global carbon cycle: A test of our knowledge of earth as a system. Science, 2000, 290: 291–296.

- Falkowski P, Scholes R J, Boyle E, et al. The global carbon cycle: A test of our knowledge of earth as a system. Science, 2000, 290: 291–296.

- Hall DO, Ojima DS, Parton WJ, Scurlock JMO (1995) Response of temperate and trop- ical grasslands to CO2 and climate change. Journal of Biogeography, 22, 537–547.

- Macías F, Arbestain M C. Soil carbon sequestration in a changing global environment. Mitig Adapt StratGl, 2010, 15: 511–529.

- Martha A S, Brian C P, Enrique R, et al. Factors affecting spatial and temporal variability in material exchange between the Southern Everglades wetlands and Florida Bay (USA). Estuar Coast Shelf Sci, 2003, 57: 757–781.

- Parr J F, Sullivan L A, Chen B, et al. Carbon bio-sequestration within the phytoliths of economic bamboo species. Glob Change Biol, 2010, 16: 2661–2667.
- Parr J F, Sullivan L A, Chen B, et al. Carbon bio-sequestration within the phytoliths of economic bamboo species. Glob Change Biol, 2010, 16: 2661–2667.
- Parr JF, Sullivan LA, Quirk R (2009) Sugarcane phytoliths: encapsulation and seques- tration of a long-lived carbon fraction. Sugarcane Technology, 11,17–21.
- Scurlock JMO, Hall DO (1998) The global carbon sink: a grassland perspective. Global Change Biology, 4, 229–233.
- Wilding L P, Brown R E, Holowaychuk N. Accessibility and properties of occluded carbon in biogenetic opal. Soil Sci, 1967, 103: 56–61.
- Zhang Y, Li C, Trettin C C, Sun G, et al. Modelling soil carbon dynamics of forested wetlands. Symposium 43. Carbon Balance of Peat lands sponsor. Inter Peat Soc, 1999.
- Zimin Li , Zhaoliang Song and Jean-Thomas Cornelis. 2014.Impact of rice cultivar and organ on elemental composition of phytoliths and the release of bio-available silicon Frontiers in plant science .Volume 5 article 529.
- Zuo X X, Lü H Y. Carbon sequestration within millet phytoliths from dry-farming of crops in China. Chin Sci Bull, 2011, 56: 3451–3456

Chapter 2

CARBON SEQUESTRATION AND CLIMATE CHANGE

**R. Murugaragavan[1], S.S.Rakesh[2],
R.Shanmugasundaram[3], S.R.Shrirangasami[4],
R.Elangovan[5] and P.T.Ramesh[6]**

1Teaching Assistant in Environmental Science, 4 Professor and Head, Department of Soils and Environment, AC&RI, Tamil Nadu Agricultural University, Madurai-625104

2 Doctoral Research Scholar in Environmental Science, Department of Environmental Sciences, Tamil Nadu Agricultural University, Coimbatore-641003

4 Assistant Professor in Agronomy, Rice Research Station, Tamil Nadu Agricultural University, Ambasamudram-627401

5 Assistant Professor in Soil Science and Agricultural Chemistry, College of Agricultural Technology, Tamil Nadu Agricultural University, Theni-625562, Tamil Nadu, India

6 Associate Professor in Environmental Science, Agricultural College and Research Institute, Tamil Nadu Agricultural University, Killikulam-628252

INTRODUCTION

Carbon sequestration is the long term storage of carbon in plants, soils, geologic formations, and the ocean. Carbon sequestration occurs both naturally and as a result of anthropogenic activities and typically refers to the storage of carbon that has the immediate potential to become CO_2 gas. Due to rapid population and industrialization concerns about climate change resulting from increased carbon dioxide concentrations in the atmosphere, considerable interest has been arrived to the possibility of increasing the rate of carbon sequestration through changes in land use pattern and afforestation and also through geoengineering techniques such as carbon capture and storage.

CARBON SOURCES AND CARBON SINKS:

Anthropogenic activities such as the burning of fossil fuels have released carbon from its long-term geologic storage as coal, petroleum, and natural gas and have delivered it to the atmosphere as carbon dioxide gas. Carbon dioxide is also released naturally, through the decomposition of plants and animals. The amount of carbon dioxide in the atmosphere has increased since the beginning of the industrial age, and this increase has been caused mainly by the burning of fossil fuels. Carbon dioxide is a very effective greenhouse gas that is, a gas that absorbs infrared radiation emitted from earth's surface. As carbon dioxide concentrations rise in the atmosphere, more infrared radiation is retained, and the average temperature of earth's lower atmosphere rises. This process is referred to as global warming (Fig.1).

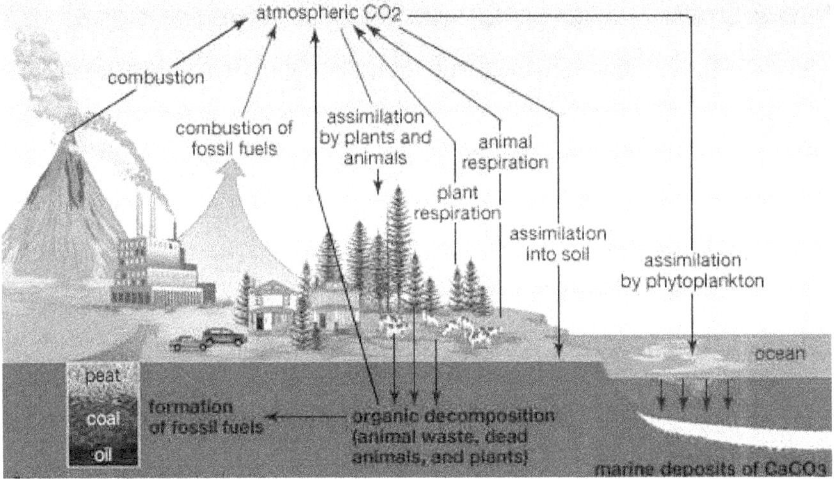

Fig.1. Sources of carbon (*Courtesy:* www.britanica.com)

Reservoirs that retain carbon and keep it from entering Earth's atmosphere are known as carbon sinks. Deforestation is a source of carbon emission into the atmosphere, but forest regrowth is a form of carbon sequestration, with the forests themselves serving as carbon sinks. Carbon is transferred naturally from the atmosphere to terrestrial carbon sinks through photosynthesis; it may be stored in aboveground biomass as well as in soils. Beyond the natural growth of plants, other terrestrial processes that sequester carbon include growth of replacement vegetation on cleared land, land-management practices that absorb carbon, and increased growth due to elevated atmospheric carbon dioxide levels and enhanced nitrogen deposition. It is important to note that carbon sequestered in soils and aboveground vegetation could be released again to the atmosphere through land use or climatic changes. For example, combustion (which is caused by fires) or decomposition (which results from microbe activity) can cause the release of carbon stored in forests to the atmosphere. Both processes join oxygen in the air with carbon stored in plant tissues to produce carbon dioxide gas (Fig.2).

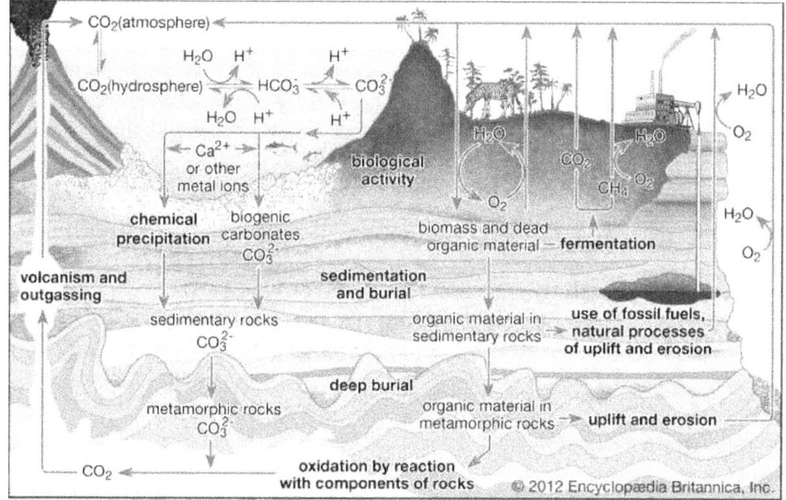

Fig.2. Sources of carbon (*Courtesy*: www.britanica.com)

CARBON CYCLE:

Carbon cycle refers to the movement of carbon, in its different forms, between the biosphere, atmosphere, oceans, and geosphere. (i) Carbon exchange between atmosphere and vegetation -Plants absorb CO$_2$ from the atmosphere during photosynthesis, and release CO$_2$ back in to the atmosphere during respiration. Another major exchange of CO$_2$ occurs between the oceans and the atmosphere. The dissolved CO$_2$ in the oceans is used by marine biota in photosynthesis. (ii)Burning of fossil fuel, coal, oil, natural gas, and gasoline are consumed by industry, power plants, and automobiles. In this process carbon directly goes to the atmosphere from its source point. (iii) Change in Land use and Land cover. Changing land use also affect carbon cycle in broad term which includes essential human activities such as agriculture, deforestation, transportation activities etc. (ETEGCC, 2004) In modern world alterations in land use and land cover and ever increasing usage of energy for the purpose of development is the focal reasons for global warming and climate change.

Carbon is transported in various forms through the atmosphere, the hydrosphere, and geologic formations. One of the primary pathways for the exchange of carbon dioxide (CO_2) takes place between the atmosphere and the oceans. The fraction of the CO_2 combines with water, forming carbonic acid (H_2CO_3) that subsequently loses hydrogen ions (H^+) to form bicarbonate (HCO_3^-) and carbonate (CO_3^{2-}) ions. Molluscs shells or mineral precipitates that form by the reaction of calcium or other metal ions with carbonate may become buried in geologic strata and eventually release CO_2 through volcanic outgassing. Carbon dioxide also exchanges through photosynthesis in plants and through respiration in animals. Dead and decaying organic matter may ferment and release CO_2 or methane (CH_4) or may be incorporated into sedimentary rock, where it is converted to fossil fuels. Burning of hydrocarbon fuels returns CO_2 and water (H_2O) to the atmosphere.

The biological and anthropogenic pathways are much faster than the geochemical pathways and, consequently, have a greater impact on the composition and temperature of the atmosphere. If the terrestrial sink becomes a significant carbon source through increased combustion and decomposition, it has the potential to add large amounts of carbon to the atmosphere and oceans. Globally, the total amount of carbon in vegetation, soil, and detritus is roughly 2,200 gigatons (1 gigaton = 1 billion tons), and it is estimated that the amount of carbon sequestered annually by terrestrial ecosystems is approximately 2.6 gigatons. The oceans themselves also accumulate carbon, and the amount found just under the surface is roughly 920 gigatons. The amount of carbon stored in the oceanic sink exceeds the amount in the atmosphere (about 760 gigatons). Of the carbon emitted to the atmosphere by human activities, only 45 percent remains in the atmosphere; about 30 percent is taken up by the oceans, and the remainder is incorporated into terrestrial ecosystems.

CARBON SEQUESTRATION AND CLIMATE CHANGE MITIGATION:

The Kyoto Protocol under the United Nations Framework Convention on Climate Change allows countries to receive credits for their carbon sequestration activities in the area of land use, land-use change, and forestry as part of their obligations under the protocol. Such activities could include afforestation (conversion of non forested land to forest), reforestation (conversion of previously forested land to forest), improved forestry or agricultural practices, and revegetation. According to the Intergovernmental Panel on Climate Change (IPCC), improved agricultural practices and forest-related mitigation activities can make a significant contribution to the removal of carbon dioxide from the atmosphere at relatively low cost. These activities could include improved crop and grazing land management for instance, more efficient fertilizer use to prevent the leaching of unused nitrates, tillage practices that minimize soil erosion, the restoration of organic soils, and the restoration of degraded lands. In addition, the preservation of existing forests, especially the rainforests of the Amazon and elsewhere, is important for the continued sequestration of carbon in those key terrestrial sinks.

CARBON CAPTURE AND STORAGE

Some policy makers, engineers, and scientists seeking to mitigate global warming have proposed new technologies of carbon sequestration. These technologies include a geoengineering proposal called carbon capture and storage (CCS). In CCS processes, carbon dioxide is first separated from other gases contained in industrial emissions. It is then compressed and transported to a location that is isolated from the atmosphere for long-term storage. Suitable storage locations might include geologic formations such as deep saline formations (sedimentary rocks whose pore spaces are saturated with water containing high concentrations of dissolved salts), depleted oil and gas reservoirs, or the deep ocean. Although CCS typically refers to the capture of carbon dioxide directly at the source of emission before it can be released into the atmosphere, it may also include techniques such as the use of scrubbing towers and "artificial trees" to remove carbon dioxide from the surrounding air.

There are many economic and technical challenges to implementing carbon capture and storage on a large scale. The IPCC has estimated that carbon capture and storage would increase the cost of electricity generation by about one to five cents per kilowatt-hour, depending on the fuel, technology, and location. Leakage of carbon from reservoirs is also a concern, but it is estimated that properly managed geological storage is very similar (that is, 66–90 percent probability) to retain 99 percent of its sequestered carbon dioxide for over 1,000 years.

CLIMATE CHANGE

Global warming and climate change refer to an increase in average global temperatures over a very long period of time. Anthropogenic activities are believed to be contributing to an increase in average global temperatures. This is caused primarily by increases in "greenhouse" gases such as Carbon Dioxide (CO_2) (Shah, 2013). Small changes in the average temperature of earth so far, can transform into large in coming hundred years. Moreover these climatic changes will have great potential to create negative impacts on environment and mankind. Therefore it is essential to mitigate climate change for advance minimization of its dangerous impacts. Current evidence suggests that to avoid the worst impacts of climate change, we should aim to limit the global average temperature rise to 2°C (35.6°F), not beyond that. This requires undertaking immediate reduction in global greenhouse gas emissions in all the sectors.

Farms emitted 6 billion tonnes of GHGs in 2011, or about 24 per cent of total global emissions. That makes the agricultural sector the world's second-largest GHG emitter after the energy sector (which includes emissions from power generation and transport). The livestock sector contributes an estimated 7,100 million tonnes of CO_2 equivalent per year, representing 14.5 per cent of human-induced greenhouse gas emissions. While there is much less methane and N_2O in the atmosphere, as outlined in Intergovernmental Panel on Climate Change (IPCC) reports, these gases have different capacities to trap heat. As a result, they are assessed using Global Warming Potential, which compares the ability of 1 kg of each gas to trap heat over a 100 year time horizon. Using this measure, methane has 25 times the warming potential of CO_2, and N_2O 298 times higher than CO_2.

The 10 countries with the largest agricultural emissions in 2011 were (in descending order): China, Brazil, United States, India, Indonesia, Russian Federation, Democratic Republic of Congo, Argentina, Myanmar, and Pakistan. Together, these countries contributed 51 per cent of global agricultural emissions. Around 39 per cent of emissions from agriculture come from only 4 countries: China, India, Brazil, and the USA. In 2011, 44 percent of agriculture-related GHG outputs occurred in Asia, followed by the Americas (25 per cent), Africa (15 per cent), Europe (12 per cent), and Oceania (4 per cent), according to FAO's data. This regional distribution was fairly constant over the last decade. In 1990, however, Asia's contribution to the global total (38 per cent) was smaller than at present, while Europe's was much larger (21 per cent).

CHANGING FACE OF GLOBAL AGRICULTURAL EMISSIONS

In 2016, the amount of atmospheric methane reached a new high of about 1,853 parts per billion (ppb) about 60 per cent comes from human activities like cattle breeding, rice agriculture and landfills. The burden of nitrous oxide was 328.9 parts per billion in 2016 about 22 per cent higher than in the pre-industrial era, mainly as a result of biomass burning and fertiliser use. From 1990 to 2010, global agricultural emissions increased 8 per cent. They are projected to increase 15 per cent above 2010 levels by 2030, when they will amount to nearly 7 billion tonnes per year. From 2012 to 2030, N2O emissions from agricultural soils are projected to increase by 3.8 per cent, from 2,114 million tonnes of CO2 equivalent to 2,195 million tonnes, which corresponds to 38.1 per cent of total agricultural emissions.

From 1990 to 2012, total emissions from other agricultural sources (such as burning of crop residues) increased from 277 million tonnes of

CO$_2$ equivalent to 301 million tonnes, corresponding to approximately 5.6 per cent of total agricultural emissions in 2010. These increases are mainly driven by population growth and changes in dietary preferences in developing economies. Agricultural emissions growth will be greatest in Asia and sub-Saharan Africa, which will account for two-thirds of the increase in overall food demand over first half of the 21st century. The production of vegetable oils and animal products with a high GHG intensity are expected to grow the most amongst agricultural outputs. Agricultural emissions from crop and livestock production grew from 4.7 billion tonnes of carbon dioxide equivalents in 2001 to over 5.3 billion tonnes in 2011, a 14 percent increase. Meanwhile, net GHG emissions due to land use change and deforestation registered a nearly 10 percent decrease over the 2001-2010 periods, averaging some 3 billion tonnes CO$_2$ eq/yr over the decade.

Agriculture is responsible for 75 per cent of global deforestation. Land use change and forestry (LUCF) caused 4 per cent of global emissions in 2010 and 14 per cent of the aggregate, global emissions from 1990 to 2011. Most of these LUCF emissions are intimately connected to agriculture, as many resulted from deforestation caused by expansion of farms into tropical forests. In contrast, forest stocks are increasing in many non-tropical regions (e.g., North America and China), due largely to forest regrowth on former farmland. Past trends imply that about 10 million square kilometres of land will be cleared by 2050 to meet food demand, leading to annual emissions of 3,000 million tonnes of CO$_2$ equivalent. Reducing land clearing to about 2 million square kilometres would reduce greenhouse gas emissions to 1,000 million tonnes of CO$_2$ equivalent per year

INDICATING FACTOR FOR CLIMATE VARIABILITY

There are several indicators on the basis of which climate variability can be measured. These

Modifications are observed by the following

- Global Green House Gas emission
- Change in precipitation pattern
- Atmospheric concentration of greenhouse gases
- Tropical Cyclonic activities
- Change in Sea level
- Increase in the occurrence of extreme events
- Change in Ocean heat
- Change in Arctic Sea ice
- Change in sea surface temperature
- Glaciers melting
- Change in Ocean Acidity
- Change in length of seasons
- Submergence of land into sea
- Length of growing season
- Climate forcing(Climate forcing refers to the amount of energy earth receive from sun and the amount of energy earth radiate back into space)
- Bird wintering ranges
- Change in Snow fall pattern
- Heat related deaths
- Change in Temperature pattern

All additional carbon dioxide in the atmosphere is increasing overall temperature of earth on day to day basis, causing global warming. It is changing climate in unpredictable ways, from floods and hurricanes to heat waves and droughts. To try and reduce the risk of global warming and extreme weather events, it is required to reduce the amount of how much fossil fuel we are burning. This isn't an easy process. In the 1997 Kyoto Protocol, it was decided that carbon emission in the atmosphere will be reduced by 5% below 1990 levels between 2008 and 2012. Several measures can be found out to reduce carbon from atmosphere and thus to reduce adverse impacts of climate change. One of the measure is carbon sequestration, which is cheap and simple as well as costly and complex. That is natural carbon sequestration and geological carbon sequestration.

CARBON SEQUESTRATION VS CLIMATE MITIGATION

- Carbon dioxide capture and sequestration could play an important role in reducing greenhouse gas emissions into the atmosphere

- It enables low-carbon electricity generation from power plants.

- As reported by INCCA (Indian Network on Climate Change Assessment) in their report, 'Green House Gas Emission2007, 38% of CO_2 emissions in India is done from electric power generation. This carbon share can be reduced by using carbon sequestration technology.

- Carbon sequestration technologies can dramatically reduce CO_2 emissions by 80-90% from power plants that burn fossil fuels. For instance - if it is Applied to a 500 MW coal fired power plant, which emits roughly 3 million tons of CO_2 per year, the amount of GHG emissions avoided (with a 90% reduction efficiency) would be equivalent to the Planting more than 62 million trees, and waiting at least 10 years for them to grow. Avoiding annual electricity related emissions from more than 300, 000 homes

- Another reason of importance is, forests, which acts as carbon sinks and store CO_2 in large amount.

The use of forest is a financially viable technique to reduce emission. It could also bring significant benefits to the local communities involved and consequently helps in reducing poverty at the same time. Forestry projects can bring social, economic, and local environmental benefits to millions of people. (Clyde, 2012)

Natural shifts in global temperatures have occurred throughout human history. The 20th century, however, has seen a rapid rise in global temperatures. Scientists attribute the temperature increase to a rise in carbon dioxide and other greenhouse gases released from the burning of fossil fuels, deforestation, agriculture and other industrial processes. Scientists refer to this phenomenon as the enhanced greenhouse effect.

The naturally occurring greenhouse effect traps the heat of the sun before it can be released back into space. This allows the Earth's surface to remain warm and habitable. Increased levels of greenhouse gases enhance the naturally occurring greenhouse effect by trapping even more of the sun's heat, resulting in a global warming effect.

The primary greenhouse gases associated with agriculture are carbon dioxide (CO_2), methane (CH_4) and nitrous oxide (N_2O). Although carbon dioxide is the most prevalent greenhouse gas in the atmosphere, nitrous oxide and methane have longer durations in the atmosphere and absorb longer wave radiation. Therefore, small quantities of methane and nitrous oxide can have significant effects on climate change.

Climate change may have beneficial as well as detrimental consequences for agriculture. Some research indicates that warmer temperatures lengthen growing seasons and increased carbon dioxide in the air results in higher yields from some crops. A warming climate and decreasing soil moisture can also result in production patterns shifting northward and an increasing need for irrigation. Changes, however, will likely vary significantly by region. Geography will play a large role in how agriculture might benefit from climate change. While projections look favorable for some areas, the potential of increased climate variability and extremes are not necessarily considered. Benefits to agriculture might be offset by an increased likelihood of heat waves, drought, severe thunderstorms and tornadoes. An increase in climate variability makes adaptation difficult for farmers.

The effects of climate on agriculture, specifically on cropping systems, pasture and grazing lands and animal management (Backlund *et al.*, 2008). The following factors are influencing the climate change

> With increased carbon dioxide and higher temperatures, the life cycle of grain and oilseed crops will likely progress more rapidly.

> The marketable yield of many horticultural crops, such as tomatoes, onions and fruits, is very likely to be more sensitive to climate change than grain and oilseed crops.

> Climate change is likely to lead to a northern migration of weeds. Many weeds respond more positively to increasing carbon dioxide than most cash crops.

> Disease pressure on crops and domestic animals will likely increase with earlier springs and warmer winters.

➢ Projected increases in temperature and a lengthening of the growing season will likely extend forage production into late fall and early spring.

➢ Climate change-induced shifts in plant species are already under way in rangelands. The establishment of perennial herbaceous species is reducing soil water availability early in the growing season.

➢ Higher temperatures will very likely reduce livestock production during the summer season, but these losses will be partially offset by warmer temperatures during the winter season (Backlund *et al.*, 2008).

Agriculture activities serve as both sources and sinks for greenhouse gases. Agriculture sinks of greenhouse gases are reservoirs of carbon that have been removed from the atmosphere through the process of biological carbon sequestration. The primary sources of greenhouse gases in agriculture are the production of nitrogen based fertilizers are the combustion of fossil fuels such as coal, gasoline, diesel fuel, natural gas and waste management. Livestock enteric fermentation, or the fermentation that takes place in the digestive systems of ruminant animals, results in methane emissions. Carbon dioxide is removed from the atmosphere and converted to organic carbon through the process of photosynthesis. As the organic carbon decomposes, it is converted back to carbon dioxide through the process of respiration. Conservation tillage, organic production, cover cropping and crop rotations can drastically increase the amount of carbon stored in soil ecosystem.

In the year 2005, agriculture accounted for from 10 to 12 percent of total global human caused emissions of greenhouse gases, according the Intergovernmental Panel on Climate Change (IPCC, 2007b). In the United States, greenhouse gas emissions from agriculture account for 8 percent of all emissions and have increased since 1990 (Congressional Research Service, 2008). In India, 68.7% percent of GHG emissions come from the energy sector, followed by agriculture, industrial processes, land-use change and forestry, and waste which contribute 19.6 percent, 6.0 percent, 3.8 and 1.9 percent relatively to GHG emissions. Greenhouse gases have varying global warming potentials; therefore climate scientists use carbon dioxide equivalents to calculate a universal measurement of greenhouse gas emissions. Agriculture's role in mitigating climate change. Several farming practices and technologies can reduce greenhouse gas emissions and prevent climate change by enhancing carbon storage in soils; preserving existing soil carbon and reducing carbon dioxide, methane and nitrous oxide emissions.

CONSERVATION TILLAGE AND COVER CROPS

Conservation tillage refers to a number of strategies and techniques for establishing crops in the residue of previous crops, which are purposely left on the soil surface. Reducing tillage reduces soil disturbance and helps mitigate the release of soil carbon into the atmosphere. Conservation tillage also improves the carbon sequestration capacity of the soil. Additional benefits of conservation tillage include improved water conservation, reduced soil erosion, reduced fuel consumption, reduced compaction increased planting and harvesting flexibility reduced labor requirements and improved soil tilth.

Organic systems of production increase soil organic matter levels through the use of composted animal manures and cover crops.

Organic cropping systems also eliminate the emissions from the production and transportation of synthetic fertilizers. Components of organic agriculture could be implemented with other sustainable farming systems, such as conservation tillage, to further increase climate change mitigation potential.

Generally, conservation farming practices that conserve moisture, improve yield potential and reduce erosion and fuel costs also increase soil carbon include direct seeding, field windbreaks, rotational grazing, perennial forage crops, reduced summer fallow and proper straw management(Alberta Agriculture and Rural Development, 2000). Using higher yielding crops or varieties and maximizing yield potential can also increase soil carbon.

LAND RESTORATION AND LAND USE CHANGES

Land restoration and land use changes that encourage the conservation and improvement of soil, water and air quality typically reduce greenhouse gas emissions. Modifications to grazing practices, such as implementing sustainable stocking rates, rotational grazing and seasonal use of rangeland, can lead to greenhouse gas reductions. Converting marginal cropland to trees or grass maximizes carbon storage on land that is less suitable for crops. Irrigation and water management Improvements in water use efficiency, through measures such as irrigation system mechanical improvements coupled with a reduction in operating hours; drip irrigation technologies; and center pivot irrigation systems, can significantly reduce the amount of water and nitrogen applied to the cropping system. This reduces greenhouse emissions of nitrous oxide and water withdrawals. Improving fertilizer efficiency through practices like precision farming using GPS tracking can reduce nitrous oxide emissions. Other strategies include the use of cover crops and manures (both green and animal), nitrogen-fixing crop rotations, composting and compost teas and integrated pest management.

METHANE CAPTURE

Large emissions of methane and nitrous oxide are attributable to livestock waste treatment, especially in dairies. Agriculture methane collection and combustion systems include covered lagoons and complete mixture and plug flow digesters. Anaerobic digestion converts animal waste to energy by capturing methane and preventing it from being released into the atmosphere. The captured methane can be used to fuel a variety of on-farm applications, as well as to generate electricity. Additional benefits include reducing odors from livestock manure and reducing labor costs associated with manure removal.

THE VALUE OF SOIL CARBON:

Potential benefits for agriculture as Mazza (2007) has remarked, "creating farm and forestry systems with strong incentives for growing soil carbon could well be at the center of climate stabilization." Thus, a new crop that farmers and ranchers may grow in the future is carbon. As with any crop, farmers and ranchers need a market for this new crop, as well as a price that will make it more profitable to grow. Value carbon from the perspective of the individual farmer and rancher, as well as society at large, is the heart of understanding the role agriculture can play in carbon sequestration and climate stabilization. To create value for offsetting greenhouse gas emissions are known as carbon taxation and cap and trade.

CARBON TAX

By taxing every ton of carbon in fossil fuels or every ton of greenhouse gas companies emit, entities that emit greenhouse gases or use carbon-based fuels will have an incentive to switch to alternative renewable fuels, invest in technology changes to use carbon based fuels more efficiently and in general adopt practices that would lower their level of greenhouse gas emissions. Thus a carbon or greenhouse gas emission tax values carbon in negative terms of tax avoidance. Those farms and ranches that emit or use less carbon intensive fuels pay a smaller tax. Farmers and ranchers use carbon-based fuels directly in the forms of petroleum and natural gas and indirectly in the forms of carbon-based fertilizers and pesticides and fuel-intensive inputs. Thus, a carbon tax could move farmers and ranchers to shift to systems of production that either eliminate the use of fossil fuels and inputs or at least improve the efficiency of their use. For the most part, carbon tax proponents have been more interested in placing greenhouse gas emission taxes on upstream producers of the original source products. This includes coal, petroleum and natural gas producers and major emitters such as large electric utilities. Nonetheless, as people work to reduce greenhouse gas emissions, the potential to place a carbon tax on sectors like agriculture may become more likely.

Benefits of a carbon tax for farmers and ranchers, a major benefit of a carbon or greenhouse gas emission tax would be the creation of a stream of tax revenue that the government could use to further induce the practice and technology changes necessary to lower greenhouse gas emissions. The current agriculture conservation programs, such as the Environmental Quality Incentive Program and the newer Conservation Stewardship Program, support improvements in soil quality and could be funded in part from emission or carbon taxes, thereby providing a revenue source to subsidize those who adopt or maintain emission reduction practices or carbon sequestration activities. Another benefit of this approach is that a tax provides a clear and stable cost to current practices. A tax also makes it easier to determine changes that will be more profitable in a new cost environment. For instance, if a concentrated animal feeding operation understood the cost of their emissions as expressed by their emission tax, it would be easier for the operation to determine alternatives to current practices that would be cost efficient. At a high enough tax rate, installing methane digesters to lower greenhouse gas emission would become economically feasible.

CAP AND TRADE:

A government-sponsored cap-and-trade system would create a new market for greenhouse gas emissions by creating a new property right; the right to emit. The market is created by a government that sets a limit or cap on total greenhouse gas emissions allowed. Companies that emit greenhouse gases are issued emission permits that allow a certain amount of emissions. Companies and groups that exceed their allowed emissions must purchase offsets from other entities that pollute less than their allowance or from entities that sequester carbon.

These exchangeable emission permits, often called allowances, are measured in tons of carbon dioxide equivalents per year. Carbon dioxide equivalents provide a common measure for all greenhouse gas emissions and are calculated by converting greenhouse gases into carbon dioxide equivalents according to their global warming potential. Over time, the government will continually lower the total level of allowances to meet an established level of acceptable total emissions. As the supply of allowances decreases, the value of the allowances will rise or fall depending on demand and on the ability of emitters to make necessary changes to reduce emissions or purchase offsets from groups more capable of reducing emissions.

OPPORTUNITIES TO REDUCE AGRICULTURAL EMISSIONS

According to estimations of the Food and Agriculture Organization of the United Nations the global demand for food is expected to increase by 3rd. 70 per cent until 2050. To ensure future food security, while reducing the impact on greenhouse gas emissions, more sustainable and resilient agriculture production is needed. Researchers from the International Institute for Applied Systems Analysis found that changes in agricultural practices and a shift away from meat and dairy could reduce 15 per cent of all agricultural methane and nitrous oxide emissions by 2050, a total of 0.8-1.4 gigatonnes of carbon dioxide equivalent per year ($GtCO_2e/y$), at an already low cost of US\$20/t CO_2e. Dietary changes in over consuming countries could contribute additional reductions of 0.6 Gt CO_2e/y, a total emissions reduction of 23%. Each year, 30 per cent of global food production is almost 1.3 billion tonnes is lost after harvest or wasted in retail and households. The direct economic cost of food wastage is \$750 billion in terms of producer prices. Environmental costs add another \$700 billion. The carbon footprint of wasted food is 3.3 GtCO2e (without emissions from land use change). The global blue water footprint of food wastage is about 250 km^3. Produced but uneaten food vainly occupies almost 1.4 billion hectares of land.

Changes in both farming practices and food demand offer big opportunities. On the supply side, crop management practices such as improved fertilizer management and conservation tillage offer the greatest reduction potential at relatively low costs. Organic agriculture also provides environmental benefits through the sequestration of atmospheric carbon in soil organic matter. Soil organic carbon stocks were 3.5 metric tonnes per hectare higher in organic than in non-organic farming systems. Organic farming systems sequestered up to 450 kg more atmospheric carbon per hectare and year through CO2 bound into soil organic matter.

CONCLUSION

Carbon sequestration is a natural process that should be enhancing by afforestation, green belt establishment, organic utilization, bio vehicles, renewable energy utilization, carbon foot print , clean development mechanism to reduce climate change impacts on natural and manmade ecosystem or man engineered ecosystem like agriculture, horticulture and pasture lands etc.,

REFERENCES

Backlund, P., et al. 2008. U.S. Climate Change Science Program and the Subcommittee on Global Change Research May 2008. The effects of climate change on agriculture, land resources, water resources, and biodiversity in the United States. *www.climatescience.gov/Library/sap/sap4-3/fi   nal-report/default.htm*

Clyde, P., (2012), importance-of-carbon-trading-in-the-global-market, slide share, available at, http://www.slideshare.net/Patricia291Clyde/importance-of-carbon-trading-in-the-global-market

Congressional Research Service. 2008. Climate Change: The Role of the U.S. Agriculture Sector. Renee Johnson. *http://fpc.state.gov/documents/organization/81931.pdf*

ETEGCC, (2004),Exploring the Environment-Global Climate Change development, Washington,available at, http://ete.cet.edu/gcc/?/about_ete/#sthash.NrKFNIZ8.dpuf

https://www.britannica.com/technology/carbon-sequestration

IPCC. 2007b. Climate Change 2007: Synthesis Report. Contribution of Working Groups I, II and III to the Fourth Assessment Report of the Intergovernmental Panel on Climate Change. [Core Writing Team, Pachauri, R.K and Reisinger, A. (eds.)]. IPCC, Geneva, Switzerland, 104 pp.

Mazza, Patrick. 2007. Growing Sustainable Biofuels Common Sense on Biofuels, Part 2.HarvestingCleanEnergyJournal(online).*http://harvestjourn al.squarespace.com/journal/2007/11/12/growing-sustainablebiofuels-producing-bioenergy-on-the-farm.html*

Shah, A., (2013), Climate Change and Global Warming Introduction, Global Issues, available at,http://www.globalissues.org/article/233/climate-change-and-global-warming-introduction

www.downtoearth.org.in

Chapter 3

MITIGATION MEASURES FOR CLIMATE CHANGE IN AGRICULTURE

S.S.Rakesh[1], R.Murugaragavan[2], B.Balamurali[3], S.Saravanakumar[4] R.Elangovan[5], P.T.Ramesh[6] and S.R.Shrirangasami[7]

1 Doctoral Research Scholar in Environmental Science, Department of Environmental Sciences, Tamil Nadu Agricultural University, Coimbatore-641003

2 Teaching Assistant in Environmental Science, Department of Soils and Environment, Agricultural College and Research Institute, TNAU, Madurai-625104

3Subject Matter Specialist in Agricultural Meteorology, Krishi Vigyan Kendra, TNAU, Papparapatty-636809, Dharmapuri District

4 Project Scientist, National Remote Sensing Centre, Hyderabad, Telungana-500037

5Assistant Professor in Soil Science and Agricultural Chemistry, College of Agricultural Technology, TNAU, Theni-625562, Tamil Nadu, India

6Associate Professor in Environmental Science, Agricultural College and Research Institute, TNAU, Killikulam-628252

7Assistant Professor in Agronomy, Rice Research Station, TNAU, Ambasamudram-627401

INTRODUCTION

The relationship between the development of environment, water, soil and crops is so complex and looks like a circle without a beginning or an end, while all others are influenced by each object. Evaporation from all water sources (rivers, irrigation canals, reservoirs, etc.) would be increased by global heating, causing a lack of water, especially in hyper-arid, arid and semi-arid regions.

High atmospheric greenhouse gas (GHG) concentrations have been caused by the burning of fossil fuels and extensive deforestation over the last few decades, and these increases in GHGs have resulted in major climate changes globally(IPCC, 2019). The world has already been warmed by climate change: from the preindustrial period (1850-1900) to the present (1998-2018), the global average land temperature has risen by 1.41°C (IPCC, 2019). This elevated global temperature has resulted in an rise in the number of hot days and nights and a decrease in the number of cold days and nights (Orlowsky and Seneviratne, 2012), as well as changes in the frequency and intensity of severe weather events such as drought and precipitation in highly regional events (IPCC, 2014). Industrialization and global population growth have led to continued deforestation and increased land demand, leading to continued high emissions of GHGs (IPCC, 2019). GHG emissions have already contributed to major changes to our ecosystems even with substantial and immediate reduction (IPCC, 2019)

A variety of variables, including climate factors, crop and land management practices, diseases and pests, and the frequency of extreme weather events, affect agricultural yields. (Pathak *et al.*, 2018; Challinor *et al.*, 2018). Wheat yields are projected to decrease by 4-6 percent with each degree of global temperature rise (Liu *et al* ., 2016), temperature rises are expected to have a similar effect on maize productivity and the areas producing 56 percent of the world's maize are expected to experience a decrease in yield by the end of the century (Bassu *et al.*, 2014; Zhao *et al.*, 2014; Pugh *et al.*, 2016).

Latitudinal distribution shifts as a consequence of climate change have already been seen by plant pathogens and pests (Bebber *et al.*, 2013), and changes in regional climate conditions are anticipated to alter plant pathogen virulence and infection rates (Vela squez *et al* ., 2018), exacerbating yield losses. It is also likely that suitable locations for planting various crops will change with changes in climatic conditions such as the number of excessively hot or cold days (Pathak *et al* ., 2018), and crop management and cultivar production will need to keep pace with these changes.

CARBON DIOXIDE CONCENTRATION

Carbon dioxide, a significant GHG, is a plant nutrient, and several crop species have shown increased yields under high CO_2 conditions (Jablonski *et al.*, 2002, Ainsworth and Long 2005), indicating that high levels of CO_2 will increase agricultural production. Experiments have shown improved yield in contained air conditions.

To maximise yield increases at higher CO_2 concentrations, improvements in crop management are required (Myers *et al.*, 2014; Tubiello *et al.*, 2000). The nutritional quality of some C_3 crop species has been decreased in terms of protein, zinc and iron under elevated CO_2, whereas C_4 crops and legumes have not been affected (Myers *et al.*, 2014, Uddling *et al* ., 2018), indicating that more agronomic adaptation and breeding would be necessary to preserve nutritional value.

Increased CO_2 concentration is correlated with resistance to drought stress (Jin *et al* . , 2018), as plants require less open stomata for gas exchange under higher CO_2 concentrations and therefore lose less water through transpiration, improving the efficiency of water usage. Short-term IPCC forecasts indicate that yield increases are anticipated in Europe and North America under several climate change scenarios due to warmer conditions that prolong the growing season at high latitudes and elevated CO_2. Increased CO_2 concentrations are less likely to compensate for climate change at lower latitudes and yield is predicted to decrease in the tropics in particular. (IPCC, 2014; Pugh *et al.*, 2016)

ABIOTIC STRESS

Long-term patterns in the occurrence of drought, heat stress and floods demonstrate regional differences in the agricultural effects of climate change. During 1964-2007, excessive heat and droughts decreased global cereal production by 9-10 percent, and droughts caused 13.7 percent greater losses from 1985 to 2007 than the estimated 6.7 percent in the previous years (1964-1984) (Lesk *et al.*, 2016). In India, warming temperatures between 1981 and 2009 decreased wheat yields by 5.2 percent. (Gupta *et al.*, 2017). In the last decade, the Hindu-Kush Himalayan area has undergone a rise in severe conditions, with agricultural yields being adversely affected by both droughts and increasing floods (Hussain *et al.*, 2016; Manzoor *et al.*, 2013). The outlook in Europe is more varied, and while trends suggest an overall decrease in wheat and barley yields of 2.5 percent and 3.8 percent , respectively, since 1989, the effects are not equally diminished, with decreases of 5 percent or greater in regions of Southern Europe (Moore and Lobell, 2015). Warming has increased the yield of fruiting vegetables in the Czech Republic by 4.9-12 percent, while root vegetable yields have decreased (Potopova *et al.*, 2017). In Russia, wheat yield per hectare and wheat production area have increased since 1980 (Di Paola *et al .*, 2018) and warming has led to an extension of the growing season in Scotland, with increased potato yields since 1960 (Gregory and Marshall 2012)

Due to climate change, climate zones are expected to move poleward, which will alter the distribution of highly productive agricultural areas, with lower yields expected at lower latitudes and higher yields expected at higher latitudes. Warmer temperatures would extend the growing season at mid and high latitudes and open up areas for the production of maize and wheat in Russia and Canada that were previously unsuitable (Pugh *et al.*, 2016). Rice production, on the other hand, is expected to worsen in the tropics, and low-latitude areas are expected to experience yield reductions. Several modelling studies have projected that production differences between most developed and developing countries will increase under a variety of climate change scenarios, where developed countries are mostly expected to benefit from climate change while developing countries are expected to undergo yield declines that will further increase the wealth gap between rich and poor nations.

BIOTIC STRESS

Atmospheric CO_2 concentrations, temperature and water availability affect pathogen virulence. By virulence and infection rates, rain, high air humidity and high soil moisture favour many plant diseases, although there are exceptions. There are optimal temperature ranges for infection for each plant-pathogen interaction, and temperature response variations have already been observed for geographically dispersed populations of *Phytophthora infestans*. (Mariette *et al.*, 2016). The virulence of *Fusarium graminearum* on wheat increased with elevated CO_2 concentrations (Vary *et al* ., 2015), while the virulence of *Peronospora manshurica* against soybean plants decreased by more than 50 percent under elevated CO_2 (Eastburn *et al.,* 2010). In plant immunity processes, where species have shown contrasting responses to humidity and temperature, a similar pattern of variability is present. Regular temperature variations of just 5oC have been shown to make potatoes more susceptible to infection with *P. infestans* (Shakya *et al.,* 2015), and temperature fluctuations also make new taxa more susceptible to invasion by bacterial communities (Saarinen et al., 2019). Given the expected increases in variability in temperature, precipitation patterns and concentrations of CO_2, changes in pathogen distribution may expose cultivars to new pathogens or strains that are more virulent.

There are some of the strategies adopted for mitigation of climate change. They are as follows

ORGANIC FARMING

Organic and agroecological solutions, as stated by Cook et al . (2016), are desperately needed to feed the planet. The path forward is sustainable, agroecological, and regenerated agriculture. These are the farming systems of this millennium and beyond. The following features are included in organic farming as an innovation: biodiverse farms foster ecological balance and protect the environment, sustainable land management, environmentally and health-friendly food production systems that provide nutritious and healthy foods, sustain food sufficiency and food sustainability, decrease carbon footprint, and produce more with fewer inputs in a vibrant agricultural systems.

But there is little acceptance of organic farming internationally, regionally, or locally (in the Philippines). There are approximately 1.6 billion hectares of total agricultural land under cultivation or exploitation for crops and livestock. In 179 countries, 50.9 M ha is organically farmed, or 3.18 percent of the 1.6 billion ha of agricultural land cultivated by more than 570 million farmers (Willer and Lernoud, 2019). Of the 2.4 million organic producers, nearly two thirds are in developing countries. Over 90% of farms are run by a person or a family that relies primarily on family labour. A large share of the world's agricultural land is occupied by family farms and about 80% of the world 's food is grown. Farms are grown using externally sourced synthetic inputs in the Asian region (ASEAN, including the Philippines), where farmers are dominated by small farms ranging from 0.5 to 3.0 ha (nearly 90 percent).

As a case study, paddy (*Oryza sativa*) cultivated by organic farming was four times more energy efficient than the traditional method. The agrochemical input (fertiliser/pesticides) accounted for 61% of the energy inputs dependent on fossil fuel and 84% of the traditional system 's cash cost of output. Just 4 calories are produced for every 1 calorie of fossil fuel energy used on a traditional farm, while 19 calories are produced on an organic farm is less energy consuming. Just 170 fossil fuel-based inputs (FFEI) were used for one tonne of paddy rice, while 844 M calories were used in traditional farming (Mendoza, 2004). The least energy intensive (lowest energy footprint), cheapest to manufacture, safe to grow (non-use of chemical pesticides; farm families are affected first), and safe to consume was organically grown rice. In addition, farmers, mostly housewives who bear the burden of where to get cash to purchase all inputs, were relieved.

Owing to crop failure, many farmers are heavily indebted (pests, floods, super typhoons, etc.). Comprehensive crop insurance in the Philippines has yet to be developed / legalised. Chemically grown rice and rice products have recently been found to have a high level of arsenic (USFDA 2016). The recent rice price increase in the Philippines clearly indicates that agriculture is in crisis (price rise of 10-15 pesos / kg in some areas declared to be a disaster region, justifying the use of their disaster funds by the local officials). Rice-growing farmers have no rice to consume. Those in the rural areas were badly affected by price jumps. The farmers had already sold their rice before it was planted. Out of 100, only 3-4 farmers have rice to eat until the next harvest season. But their living standards have been changed by organic farmers. They actually have enough rice to eat, are no longer indebted, are happier, and are willing to send their kids to school (Mendoza 2004).

Farmers choose their own locally-adapted and target-market seeds, often coloured rice (red, black rice) as requested by consumers. Scheduling the preparation of land in such a way that ample lead time is provided for the decomposition of crop / weed residues, that is, 3-4 weeks before transplanting. Planting at the correct time to reduce the infestation of pests and diseases. Planting of three types of sugarcane varieties for early, normal, and late ripening (*Saccharum officinarum*). Farmers know when and how to plant seeds by season (wet, dry season) and location (soil type, topography) (locally adapted, G X E interaction) to obtain the genetic potential of high seed yields (inbred or hybrid). Implementing soil quality management practises such as crop / weed residue recycling by spreading it uniformly on the farm by its decomposition after harvesting. Sugarcane planters prevent the burning of sugarcane trash and either carry out trash shredding through a mechanical shredder or incorporate disc harrows drawn by the tractor repeatedly into the soil. In terms of labour and cost of production, farmers prepare their own compost / organic fertiliser and other preparations, thus saving a lot. Farmers on the island of Mindanao in Cagayan de Oro process their organically grown black rice, which is best known for its health and export market benefits.

Through the processing of cacao beans into tablea, cacao farmers in Kidapawan City , North Cotabato, increased their income as cacao bean prices increased from PHP 120 / kg to PHP 600 / kg. Coffee farmers in Batangas grind their coffee beans at PHP 216 / kg and market them as kapeng barako. Dried coffee beans, on the other hand, are only sold for PHP 80 / kg. Many organic farmers sell their farm products directly and some have their own restaurant to further increase the value of their organic farm produce through "fine dining" at PHP 1200–PHP 1800 per head.

ECOLOGICAL TAX

An ecological tax is a groundbreaking policy to promote organic agriculture. It should be introduced by environmentalists and proponents of organic agriculture. The Landicho et al . (2016) research validated the understanding of the benefits of cultivating and eating organic produce by farmers. The findings obtained substantiate that knowledge is being generated at the level of the grassroots. Farmers themselves have explicitly explained that organic farming activities are environmentally friendly, reducing air pollution, greenhouse gases emissions, thereby contributing as well to a healthy society. The sustainable ecosystem is rooted in healthy soils of organically grown crops. For both producers and consumers, this ensures safe , nutritious, and organic foods. Farmers' respondents also confirmed that organic farming resulted in higher sales due to lower production costs and a lower chance of crop failure. Organic farming practises increase the valuation of global ecosystem resources from a broader perspective, namely, the provision, regulation and promotion (Fan *et al.*, 2016).

INTERCROPPING

A six-fruit tree-intercrops under coconut (*Cocos nucifera*) value chain assessment was carried out (Mendoza *et al.*, 2016). Coconut revenue alone is very poor. But coconut offers many ecosystem services, including coffee (*Coffea sp.*), cacao (*Theobroma cacao*), mangosteen (*Garcinia mangostana*), durian (*Durio zibethinus*), banana (*Musa sp.*), and even rubber (*Hevea brasiliensis*) trees, a microclimate modifier that provides a "shading effect" beneficial to understory fruit trees. Intercropping increased the net revenue using monoculture coco–copra revenue as base revenue at P 18,251/ha as follows: coconut + banana increase the revenues 12.56 X; coconut + mango (*Mangifera indica*) 9.37X; coconut + durian 7.83X; coconut + mangosteen 5.34X; coconut + cacao 2.83X; and coconut + coffee 2.36X. Income increases progressively as the product move across the value chain.

In coconut + mango, sales multiply 10 times (coffee, cacao) and as high as 32 times before the insect and disease issues set in. It is imprecise, however, to declare that one crop is more profitable than the others. Cacao dried beans are sold at PHP 120 / kg higher than PHP 85 / kg coffee beans. The gross earnings could go as high as PHP 8.4 million worth of coffee ha^{-1}, when Starbucks sells coffee. Cacao made into chocolate could yield millions (more than PHP 2 million ha^{-1}). For durian and mangosteen, sales of mangosteen and/or durian from farm gates are comparatively low. But selling them raises revenue 2.15 times (durian) to 3.5 times (mangosteen) by retailing them.

WEATHER BASED AGRO ADVISORIES

Weather based Agro Advisory Service is a farming response tool that enables farmers to obtain timely weather-based agro advice to make the required decisions for the next few days of farm operations. For closer spatial size, high performance computing server, faster internet access, short messaging service (SMS) and web cum mobile application, the AAS needs previous and expected weather data of 6 days from the current date. With the support of technocrats, weather-based agro-advisory has been developed for multiple crops and various stages of crop growth and integrated into the database (Dheebakaran *et al*., 2020). Farmers must register their mobile number, unique to the crop process, for advice on their own crop. Every day, for past and future weather, each block 's weather scenario will be developed separately and the AAS module matches the scenario, crop, point, and advisory and sends the chosen advisory to the farmers mobile as SMS. Via a web portal or mobile app, farmers may change their crops and the sowing date.

In India, the Meteorological Department of India (IMD) is introducing the GKMS programme at 127 centres in all states. Approximately 11 lakh farmers are currently benefiting from this service, which has not only helped farmers increase crop yield, but also reduced losses due to bad weather. In Tamil Nadu Automated Agro Advisory Service (AAS) is mobilized by Tamil Nadu Agricultural University all over the state.

AGROECOTOURISM

The type of tourism that capitalises on rural culture as a tourist attraction is "agro-tourism." As a potential income and work generating operation, it has acquired a new dimension. A core aspect of environmentally and socially responsible tourism is the symbiosis between tourism and agriculture that can be found in agro-tourism. Agroecotourism provides an opportunity to experience the true enchanting and authentic interaction with rural life, taste genuine local food, and get familiar with the various farming tasks. It is perhaps "a welcome escape from the everyday hectic life in the quiet rural area and to relax and revitalise in the pure natural environment, surrounded by beautiful scenery." Via investments leading to energy and water production, waste reduction, protection of biodiversity and cultural heritage, and strengthening relations with local communities, agro-ecological tourism will contribute to the transition to a green economy (Barbuddhe and Singh 2014). Agroecotourism in the Philippines (Costales Farm, Quezon; Penalosa farm in Victorias, Negros Occidental) increased the market value of their organic products through their package tour as a training venue, as a direct marketing for their products, buying organic products like seedlings when the agroecotourists go home.

CONCLUSION

Climate change is one of the serious threat to the entire globe which involves multiple negative consequences such as global warming, alteration in crop patterns and yields, monsoon failure, drought, famine, flood events, glacier and polar ice cap melting, ocean acidification, habitat loss, loss of biodiversity which needs to be mitigated as soon as possible which ensures the globe for human inhabitation for future generations and conservation of the flora and fauna is essential for sustainability.

REFERENCES

Ainsworth EA, Long SP: What have we learned from 15 years of free-air CO2 enrichment (FACE)? A meta-analytic review of the responses of photosynthesis, canopy properties and plant production to rising CO2. New Phytol 2005, 165:351-372.

Barbuddhe, S.B. and Singh, N.P., 2014. Agro-eco-tourism: a new dimension to agriculture.

Bassu, S., Brisson, N., Durand, J.L., Boote, K., Lizaso, J., Jones, J.W., Rosenzweig, C., Ruane, A.C., Adam, M., Baron, C. and Basso, B., 2014. How do various maize crop models vary in their responses to climate change factors?. *Global change biology, 20*(7), pp.2301-2320.

Bebber, D.P., Ramotowski, M.A. and Gurr, S.J., 2013. Crop pests and pathogens move polewards in a warming world. *Nature climate change, 3*(11), pp.985-988.

Challinor, A.J., Müller, C., Asseng, S., Deva, C., Nicklin, K.J., Wallach, D., Vanuytrecht, E., Whitfield, S., Ramirez-Villegas, J. and Koehler, A.K., 2018. Improving the use of crop models for risk assessment and climate change adaptation. *Agricultural systems, 159,* pp.296-306.

Cook, J., Oreskes, N., Doran, P.T., Anderegg, W.R., Verheggen, B., Maibach, E.W., Carlton, J.S., Lewandowsky, S., Skuce, A.G., Green, S.A. and Nuccitelli, D., 2016. Consensus on consensus: a synthesis of consensus estimates on human-caused global warming. *Environmental Research Letters, 11*(4), p.048002.

Dheebakaran, G., Panneerselvam, S., Geethalakshmi, V. and Kokilavani, S., 2020. Weather Based Automated Agro Advisories: An Option to Improve Sustainability in Farming Under Climate and Weather Vagaries. In *Global Climate Change and Environmental Policy* (pp. 329-349). Springer, Singapore.

Di Paola A, Caporaso L, Di Paola F, Bombelli A, Vasenev I, Nesterova OV, Castaldi S, Valentini R: The expansion of wheat thermal suitability of Russia in response to climate change. Land Use Policy 2018, 78:70-77.

Eastburn DM, Degennaro MM, Delucia EH, Dermody O, Mcelrone AJ: Elevated atmospheric carbon dioxide and ozone alter soybean diseases at SoyFACE. Glob Change Biol 2010, 16:320-330.

Fan, J., Wu, L., Zhang, F., Xiang, Y. and Zheng, J., 2016. Climate change effects on reference crop evapotranspiration across different climatic zones of China during 1956–2015. *Journal of Hydrology, 542,* pp.923-937.

Gregory PJ, Marshall B: Attribution of climate change: a methodology to estimate the potential contribution to increases in potato yield in Scotland since 1960. Glob Change Biol 2012, 18:1372-1388.

Gupta R, Somanathan E, Dey S: Global warming and local air pollution have reduced wheat yields in India. Clim Change 2017, 140:593-604.

Hussain A, Rasul G, Mahapatra B, Tuladhar S: Household food security in the face of climate change in the Hindu-Kush Himalayan region. Food Secur 2016, 8:921-937.

IPCC: Climate Change 2014: Synthesis Report. Contribution of Working Groups I, II and III to the Fifth Assessment Report of the Intergovernmental Panel on Climate Change. Intergovernmental Panel on Climate Change. 2015.

IPCC: Climate Change and Land: an IPCC Special Report on Climate Change, Desertification, Land Degradation, Sustainable Land Management, Food Security, and Greenhouse Gas Fluxes in Terrestrial Ecosystems (SRCCL). 2019.

Jablonski LM, Wang X, Curtis PS: Plant reproduction under elevated CO2 conditions: a meta-analysis of reports on 79 crop and wild species. New Phytol 2002, 156:9-26. 31.

Jin Z, Ainsworth EA, Leakey ADB, Lobell DB: Increasing drought and diminishing benefits of elevated carbon dioxide for soybean yields across the US Midwest. Glob Change Biol 2018, 24:e522-e533.

Landicho, L., Paelmo, R., Cabahug, R., de Luna, C., Visco, R. and Tolentino, L., 2016. Climate change adaptation strategies of smallholder agroforestry farmers in the Philippines. *Journal of Environmental Science and Management, 19*(1).

Lesk C, Rowhani P, Ramankutty N: Influence of extreme weather disasters on global crop production. Nature 2016, 529:84-87.

Liu, Z., Davis, S.J., Feng, K., Hubacek, K., Liang, S., Anadon, L.D., Chen, B., Liu, J., Yan, J. and Guan, D., 2016. Targeted opportunities to address the climate–trade dilemma in China. *Nature Climate Change, 6*(2), pp.201-206.

Long SP, Ainsworth EA, Leakey ADB: Food for thought: lower-than-expected crop yield stimulation with rising CO_2 concentrations. Science 2006, 312:5.

Manzoor M, Bibi S, Manzoor M, Jabeen R: Historical analysis of flood information and impacts assessment and associated response in Pakistan (1947-2011). Res J Environ Earth Sci 2013, 5:139-146.

Mariette N, Androdias A, Mabon R, Corbie` re R, Marquer B, Montarry J, Andrivon D: Local adaptation to temperature in populations and clonal lineages of the Irish potato famine pathogen *Phytophthora infestans*. Ecol Evol 2016, 6:6320-6331. 53.

Mendoza, A.R., Corvo, F., Gómez, A. and Gómez, J., 2004. Influence of the corrosion products of copper on its atmospheric corrosion kinetics in tropical climate. *Corrosion Science, 46(5)*, pp.1189-1200.

Mendoza, P.A., Mizukami, N., Ikeda, K., Clark, M.P., Gutmann, E.D., Arnold, J.R., Brekke, L.D. and Rajagopalan, B., 2016. Effects of different regional climate model resolution and forcing scales on projected hydrologic changes. *Journal of Hydrology, 541*, pp.1003-1019.

Moore FC, Lobell DB: The fingerprint of climate trends on European crop yields. Proc Natl Acad Sci U S A 2015, 112:2670- 2675.

Myers SS, Zanobetti A, Kloog I, Huybers P, Leakey ADB, Bloom AJ, Carlisle E, Dietterich LH, Fitzgerald G, Hasegawa T et al.: Increasing CO2 threatens human nutrition. Nature 2014, 510:139-142.

Orlowsky, B., & Seneviratne, S. I. (2012). Global changes in extreme events: regional and seasonal dimension. *Climatic Change, 110(3-4)*, 669-696.

Pathak TB, Maskey ML, Dahlberg JA, Kearns F, Bali KM, Zaccaria D: Climate change trends and impacts on california agriculture: a detailed review. Agronomy 2018, 8:25.

Potopova V, Zahradnıcek P, St epanek P, Turkott L, Farda A, Soukup J: The impacts of key adverse weather events on the field-grown vegetable yield variability in the Czech Republic from 1961 to 2014. Int J Climatol 2017, 37:1648-1664.

Pugh, T.A.M., Müller, C., Elliott, J., Deryng, D., Folberth, C., Olin, S., Schmid, E. and Arneth, A., 2016. Climate analogues suggest limited potential for intensification of production on current croplands under climate change. *Nature Communications, 7*(1), pp.1-8.

Saarinen K, Lindstro¨m L, Ketola T: Invasion triple trouble: environmental fluctuations, fluctuation-adapted invaders and fluctuation-mal-adapted communities all govern invasion success. BMC Evol Biol 2019, 19:42.

Seneviratne SI, Nicholls N, Easterling D, Goodess CM, Kanae S, Kossin J, Luo Y, Marengo J, Innes KM, Rahimi M et al.: Changes in climate extremes and their impacts on the natural physical environment. Managing the Risks of Extreme Events and Disasters to Advance Climate Change Adaptation: Special Report of the Intergovernmental Panel on Climate Change. Cambridge University Press; 2012:109-230.

Shakya SK, Goss EM, Dufault NS, van Bruggen AHC: Potential effects of diurnal temperature oscillations on potato late blight with special reference to climate change. Phytopathology 2014, 105:230-238.

Tubiello FN, Donatelli M, Rosenzweig C, Stockle CO: Effects of climate change and elevated CO2 on cropping systems: model predictions at twoItalian locations. EurJ Agron 2000,13:179-189.

Uddling J, Broberg MC, Feng Z, Pleijel H: Crop quality under rising atmospheric CO2. Curr Opin Plant Biol 2018, 45:262-267.

USFDA,(2016)

http://www.fda.gov/food/foodborneillnesscontaminants/ metals/ucm319870.htm)

Vary Z, Mullins E, McElwain JC, Doohan FM: The severity of wheat diseases increases when plants and pathogens are acclimatized to elevated carbon dioxide. Glob Change Biol 2015, 21:2661-2669. 54.

Velasquez, A.C., Castroverde, C.D.M. and He, S.Y., 2018. Plant–pathogen warfare under changing climate conditions. *Current Biology, 28*(10), pp.R619-R634.

Willer, H. and Lernoud, J., 2019. *The world of organic agriculture. Statistics and emerging trends 2019* (pp. 1-336). Research Institute of Organic Agriculture FiBL and IFOAM Organics International.

Zhao, L., Lee, X., Smith, R.B. and Oleson, K., 2014. Strong contributions of local background climate to urban heat islands. *Nature, 511*(7508), pp.216-219.

Chapter 4

CLIMATE CHANGE RISK MANAGEMENT USING REMOTE SENSING AND GEO-SPATIAL TECHNOLOGIES

J. Ramachandran[1] M. Rajeswari[2] and K. Arunadevi[3]

[1]Teaching Assistant, [2]Professor and Head, Department of Agricultural Engineering, Agricultural College and Research Institute, Tamil Nadu Agricultural University, Madurai 624 105
[3]Assistant Professor (SWCE), Department of Soil and Water Conservation Engineering, AEC&RI, Kumulur 621712

INTRODUCTION

Climate change disasters have become an issue of growing concern throughout the world. In particular, the relationship between climate change adaptation and disaster risk reduction has received significant attention (Mercer, 2010; Schipper, 2009; Schipper and Pelling, 2006; Thomalla et al., 2006), leading to calls for greater policy connections between the two domains. Aiming to address some of these issues amid growing concern about climate-related disasters, the Intergovernmental Panel on Climate Change produced a Special Report, Managing the Risks of Extreme Events and Disasters to Advance Climate Change Adaptation (IPCC, 2012). Currently emerging local responses to climate change and disaster risk include the integration of climate risk information into disaster planning and community-based adaptation, which includes a strong local participation component (Cutter et al., 2012). The frequency and magnitude of disasters menacing large population living in diverse environments is increasing in recent years across the world. These disasters also have far-reaching implications on sustainable development through social, economic and environmental impacts. It is highly imperative to develop effective tools for disaster management. Remote sensing systems have been playing a great role in disaster management in such areas as flooding, cyclones, drought, earthquake and tsunami. Satellite remote sensing is largely adopted due to its cost effectiveness, short temporal orbiting and large area of coverage. Disaster Management can be defined as the organization and management of resources and responsibilities for dealing with all humanitarian aspects of emergencies, in particular preparedness, response and recovery in order to lessen the impact of disasters. Remote sensing technologies have been used in disaster management especially during the preparedness/warning and response/monitoring stages. Using Geographical Information System (GIS) technology and by doing spatial analysis, it is possible to design highly accurate interactive simulation systems that promote a more broad understanding of various disasters, their outcomes and the damage inflicted on a given area. GIS techniques act as a decision support tool because all disasters are spatial in nature. With the help of the different GIS layers, decision making can be made possible. This chapter highlights application of remote sensing methodologies and GIS techniques in the event of disaster management with few case studies.

FLOOD

The flood situation is extreme when the water level of a river touches or crosses the Highest Flood Level (HFL) recorded at any forecasting site so far, read the Central Water Commission definition. For example, two rivers in Maharashtra — Warna and Krishna — and one in Karnataka — Dudhganga — not only crossed danger marks but also the HFL of 2005 on August 8 and 9. Both Maharashtra and Karnataka suffered damages in devastating deluges recently. Climate change has impacted India severely. Extreme rainfall events and widespread floods have increased manifold over the last several decades. States like Rajasthan, Maharashtra, Karnataka, and Gujarat received 36, 30, 22, and 31 per cent more rainfall than normal between June 1 and September 18, 2019. This is the highest among the big states of India. In Kerala, it looks like climate-change-induced floods are becoming an annual affair. Following heavy rains, the state which once boasted consistent rains during the monsoon season is now grappling with devastating floods in the month of August. In this situation, a remote sensing methodology helps in monitoring and impact studies of flood prone areas. The Bhuvan – India's National Geo-portal is amazingly serving in identifying the flood affected areas.

The following is the image (Fig. 1) showing the flood duration map of Kerala floods during 11-21 August 2018. The flood prone area in Mumbai is also depicted in Fig. 2. These are valid sources in preparing mitigation strategies and damage assessment.

The possible application of remote sensing in various stages of flood is given in table 1.

Table 1. Applications at Various Stages of Flood.

Response	Preparedness	Mitigation	Recovery
Flood mapping; evacuation planning; damage assessment	Flood detection; early warning; rainfall mapping	Identifying and mapping flood-prone areas, delineating flood-plains, land-use mapping	Damage assessment. Spatial planning

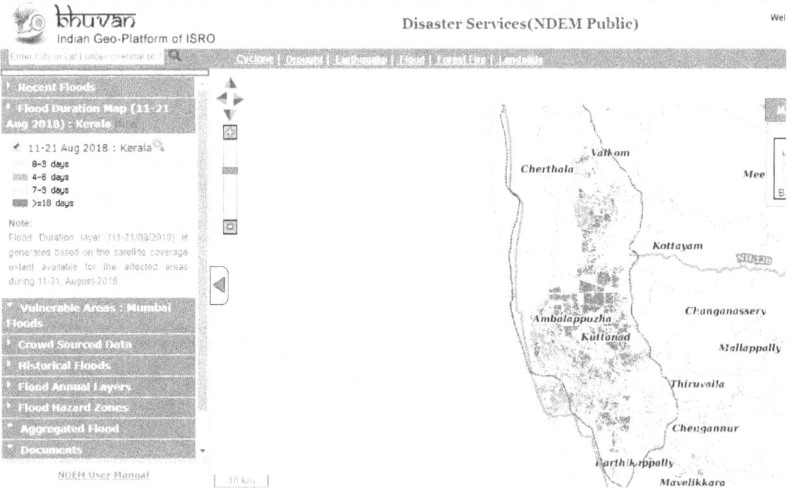

Figure 1. Flood Duration Map of Kerala (Source: Bhuvan, Indian Geo-Platform of ISRO)

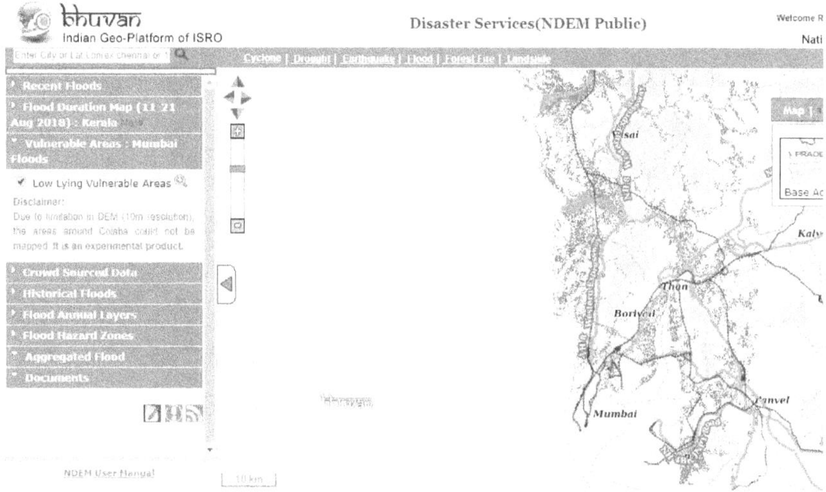

Figure 2. Flood Prone Areas Map of Mumbai

(Source: Bhuvan, Indian Geo-Platform of ISRO)

CYCLONE

Cyclones are strong spiralling winds characterized by low pressure and numerous thunderstorms. In the past many years, cyclones have been a part of the country, the southern part of the country to be precise. Cyclones generally occur in the month of May-June and October-November, with a primary peak in November and secondary peak in May. Although cyclones affect the entire coast of India, the East Coast is more prone compared to the West Coast. A number of structural and non-structural measures can be adopted for effective management of cyclones.

The structural measures include construction of cyclone shelters, construction of cyclone resistant buildings, road links, culverts, bridges, canals, drains, saline embankments, surface water tanks, communication and power transmission networks etc. Non-structural measures involve early warning dissemination systems, management of coastal zones, awareness generation and disaster risk management and capacity building of all the stakeholders. Remote sensing and GIS techniques plays a vital role in cyclone warning and mitigation. The cyclone warning in India is given by Indian Meteorological Department. The cyclone intensity and direction of movement in predicted by the department. The following is the image (Fig. 3) of cyclone prediction dated 19.12.2018.

Cyclone Gaja made landfall over Tamil Nadu coast on 16th November 2018 at 1:45 AM, wreaking havoc as it passed the coastal districts of Tamil Nadu. Cyclone Gaja made its landfall with wind speed of 120 – 140 kph, leaving behind a huge trail of destruction in the districts of Nagapattinam, Cuddalore, Thanjavur, Tiruvarur, Pudukottai, Dindigul, Trichy, Karur, Sivagangai, Ramnad, and Karaikal in the Union territory of Pondicherry. According to the Government record, 45 people were killed and around 250,000 people were displaced from their homes on the coasts of Tamil Nadu and Puducherry. The rainfall accumulation forecast given by the weather company, India is shown in Fig. 4.

Figure 3. Cyclone Warning System (Source: IMD, New Delhi).

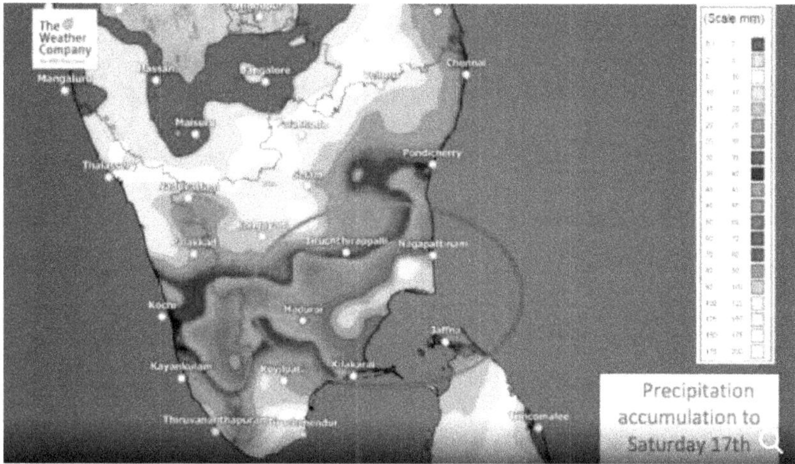

Figure 4. Rainfall Accumulation Forecast (Source: The Weather Company, India)

The Bhuvan – National Geoportal has given the severity zones of Gaja Cyclone as depicted in Fig. 5. This helps in planning and adopting mitigation measures in high impact zones. The possible application of remote sensing in various stages of cyclone is given in table 2.

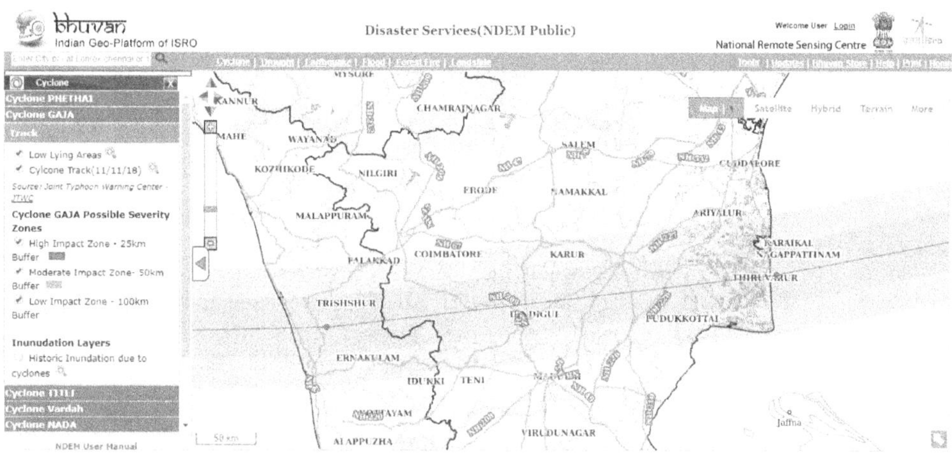

Figure 5. Cyclone GAJA Possible Severity Zones (Source: Bhuvan, Indian Geo-Platform of ISRO)

Table 2. Applications at Various Stages of Cyclone.

Response	Preparedness	Mitigation	Recovery
Impact Assessment, Identifying routes to escape, Crisis Mapping, Regular monitoring of cyclones and Storm surge predictions.	Early warning signs, long range climate modeling	Vulnerability Analysis and Risk Modeling	Damage assessment; spatial planning

DROUGHT

More than 44 per cent of India's areas were under various degrees of drought conditions (abnormally dry to exceptionally dry) as of June 10, 2019 — that is nearly 11 percentage point over a year ago, according to the Drought Early Warning System (DEWS).India has witnessed the second-driest pre-monsoon season in the last 65 years.

The country received 99 millimeters (mm) rainfall between March and May — with 23 per cent below the normal rainfall during this time of the year — showed the latest data from the India Meteorological Department (IMD). The deficit is prevalent across the country with south India having a deficit of 47 per cent, followed by northwest India (30 per cent), central India (18 per cent), and east and northeast (14 per cent) (Source: www.downtoearth.org.in).

The meteorological indicator like Standard Precipitation Index can be used in drought prediction. The spectral indicator like Normalized Difference Vegetation Index, Normalized Difference Water Index are also useful in identifying the drought prone areas. The Bhuvan geo-portal provides Soil Moisture Index for all over India which helps in drought management and mitigation.

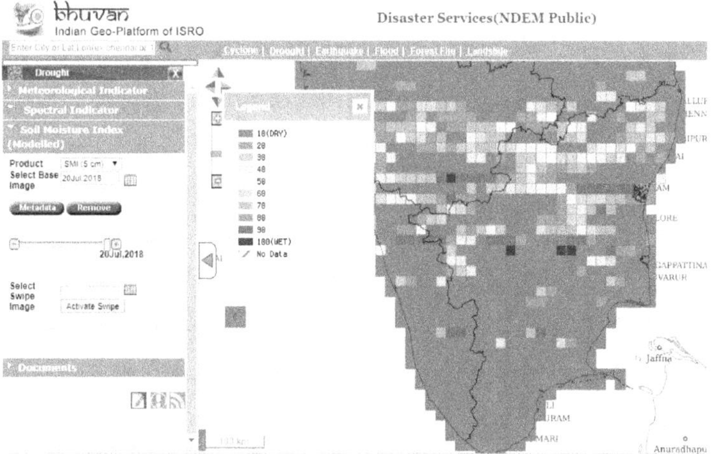

Figure 5. Soil Moisture Index (Modelled) (Source: Bhuvan, Indian Geo-Platform of ISRO)

The possible application of remote sensing in various stages of drought is given in table 3.

Table 3. Applications at Various Stages of Drought.

Response	Preparedness	Mitigation	Recovery
Assessing the extent of damage,	Weather forecasting; vegetation monitoring;	Risk modelling; vulnerability analysis;	Informing drought mitigation

monitoring vegetation	crop water requirement mapping; early warning	land and water management planning	

EARTHQUAKE

Earthquake - prone areas of the country have been identified on the basis of scientific inputs relating to seismicity, earthquakes occurred in the past and tectonic setup of the region. Based on these inputs, Bureau of Indian Standards [IS 1893 (Part I):2002], has grouped the country into four seismic zones, viz. Zone II, III, IV and V. Of these, Zone V is seismically the most active region, while zone II is the least. Broadly, Zone - V comprises entire northeastern India, parts of Jammu and Kashmir, Himachal Pradesh, Uttaranchal, Rann of Kutch in Gujarat, part of North Bihar and Andaman & Nicobar Islands. Zone - IV covers remaining parts of Jammu and Kashmir and Himachal Pradesh, National Capital Territory (NCT) of Delhi, Sikkim, Northern Parts of Uttar Pradesh, Bihar and West Bengal, parts of Gujarat and small portions of Maharashtra near the west coast and Rajasthan. Zone – III comprises Kerala, Goa, Lakshadweep islands, remaining parts of Uttar Pradesh, Gujarat and West Bengal, Parts of Punjab, Rajasthan, Madhya Pradesh, Bihar, Jharkhand, Chhattisgarh, Maharashtra, Orissa, Andhra Pradesh, Tamilnadu and Karnataka. Zone - II covers remaining parts of country. This kind of map is mainly used by the Department of Disaster Management of the different state governments in the country. The possible application of remote sensing in various stages of earthquake is given in table 4.

Table 4. Applications at Various Stages of Earthquake.

Response	Preparedness	Mitigation	Recovery
Identifying escape routes, planning routes for search and rescue	Measuring strain accumulation.	Hazard mapping and assessment of building stock	Damage assessment; identifying sites for rehabilitation

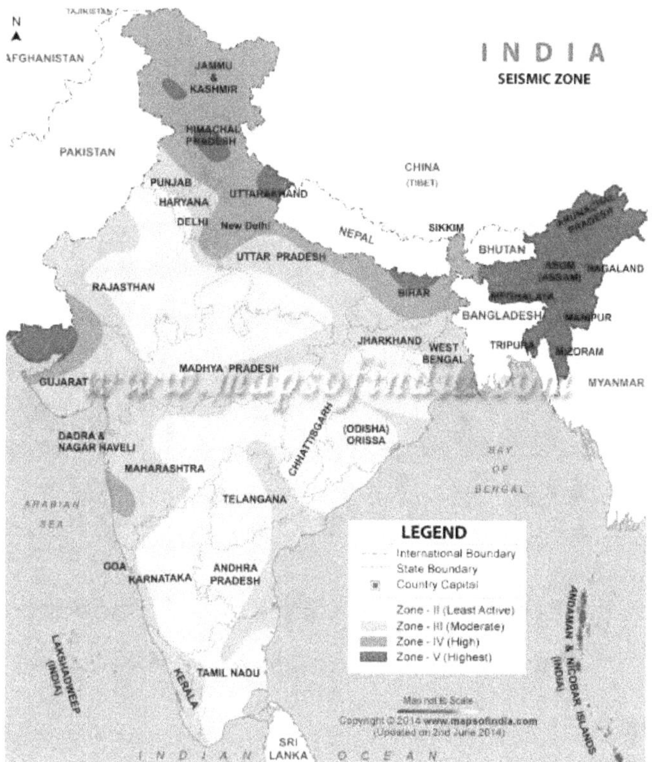

Figure 6. Seismic Zone map of India (www.mapsofindia.com)

This map helps them in planning for a natural disaster like earthquake. An Indian seismic zoning map assists one in identifying the lowest, moderate as well as highest hazardous or earthquake prone areas in India. The damage assessment after earthquake and pre and post event land use/cover changes can be studied through remote sensing and GIS (Fig. 7a & 7b).

Figure 7a. Pre Event Image of Kathmandu Earthquake (Source: Bhuvan, Indian Geo-Platform of ISRO)

Figure 7b. Pre Event Image of Kathmandu Earthquake (Source: Bhuvan, Indian Geo-Platform of ISRO)

CONCLUSION

India is prone to many natural disasters like floods, cyclones, earthquakes, drought, etc. Climate change will cause more hazards, and it will be more severe. Coupled with persistent poverty and governance failures around the world, this means disaster risk is likely to increase. Satellites provide synoptic observations of the natural disasters at regular intervals that helps in better planning and management of disasters. In order to better understand the risks due to such disasters, it is necessary to integrate satellite and field based observations and to work towards risk reduction principles. Various applications remote sensing and GIS techniques in disaster management is briefed in this chapter.

REFERENCES

1. Cutter, S., Osman-Elasha, B., Campbell, J., Cheong, S.-M., McCormick, S., Pulwarty, R., Supratid, S. and Ziervogel, G. 2012. Managing the risks from climate extremes at the local level. In A Special Report of Working Groups I and II of the Intergovernmental Panel on Climate Change (IPCC). C. B. Field, V. Barros, T. F. Stocker, D. Quin, D. J. Dokken, et al. (eds.). Cambridge, UK, and New York, NY, USA: Cambridge University Press. pp. 291–338.

2. IPCC. 2012. Managing the Risks of Extreme Events and Disasters to Advance Climate Change Adaptation. A Special Report of the Intergovernmental Panel on Climate Change Working Groups I and II (Field, C.B., V. Barros, T.F. Stocker, D. Qin, D.J. Dokken, K.L. Ebi, M.D. Mastrandrea, K.J. Mach, G.-K. Plattner, S.K. Allen, M. Tignor, and P.M. Midgley, eds.). Cambridge, UK, and New York: Cambridge University Press. http://ipcc-wg2.gov/SREX/.

3. Mercer, J. 2010. Disaster risk reduction or climate change adaptation: Are we reinventing the wheel? Journal of International Development, 22(2). 247–64. DOI:10.1002/jid.1677.

4. Schipper, L. and Pelling, M. 2006. Disaster risk, climate change and international development: scope for, and challenges to,

integration. Disasters, 30(1). 19–38. DOI:10.1111/j.1467-9523.2006.00304.x.

5. Schipper, E. L. F. 2009. Meeting at the crossroads?: Exploring the linkages between climate change adaptation and disaster risk reduction. Climate and Development, 1(1). 16–30. DOI:10.3763/cdev.2009.0004.

6. Thomalla, F., Downing, T., Spanger-Siegfried, E., Han, G. and Rockström, J. 2006. Reducing hazard vulnerability: towards a common approach between disaster risk reduction and climate adaptation. Disasters, 30(1). 39–48. DOI:10.1111/j.1467-9523.2006.00305.x.

Chapter 5

BIOCHAR IN CARBON SEQUESTRATION

P. Christy Nirmala Mary

Associate Professor (SS&AC), Dept. of Soils and Environment, Agricultural College and Research Institute, TNAU, Madurai

INTRODUCTION

Soils store three times more carbon than exists in the atmosphere. Plants absorb atmospheric carbon during photosynthesis, so the return of plant residues into the soil contributing to soil carbon. While much of this carbon ultimately returns to the atmosphere as soil microbes decompose carbon based plant biomass and release carbon dioxide, soil carbon stores can increase if the rate of carbon inputs exceeds the rate of microbial decomposition. The term "carbon sequestration" is used to describe both natural and deliberate processes by which CO_2 is either removed from the atmosphere or diverted from emission sources and stored in the ocean, terrestrial environments (vegetation, soils, and sediments), and geologic formations. Before human-caused CO_2 emissions began, the natural processes that make up the global "carbon cycle" (fig. 1) maintained a near balance between the uptake of CO_2 and its release back to the atmosphere. However, existing CO_2 uptake mechanisms (sometimes called CO_2 or carbon "sinks") are insufficient to offset the accelerating pace of emissions related to human activities.

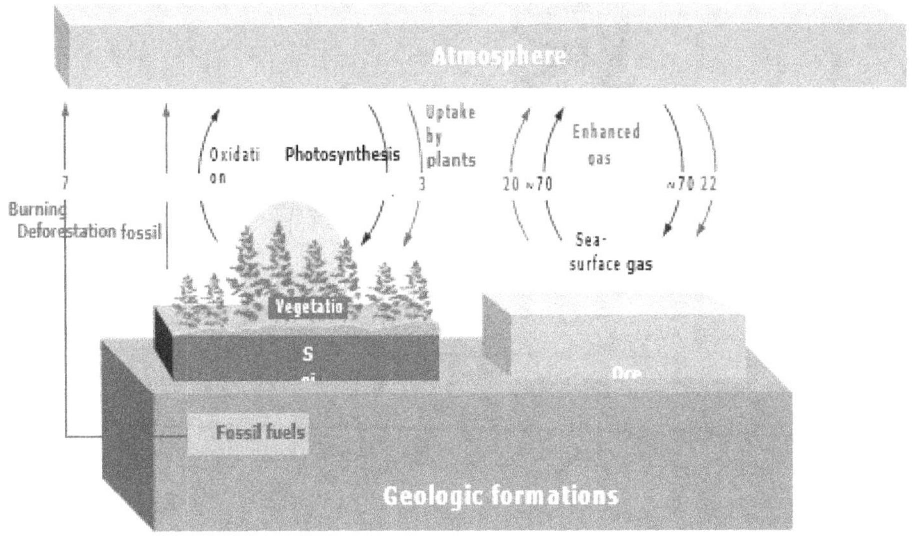

Fig. 1 Global Carbon Cycle

Three of the biggest challenges of the twenty-first century are the need to nearly double food production by 2050, to adapt and build resilience to a more and more challenging climatic environment, and to simultaneously achieve a substantial reduction in atmospheric greenhouse gas concentrations. The surge of interest in climate-smart agriculture, which focuses on solutions to the three challenges, has sparked curiosity in using biochar as a tool to fight climate change while also improving soil fertility. Biochar systems are particularly relevant in developing country contexts and could be leveraged to address global challenges associated with food production and climate change. However the potential effects of biochar application to soils are diverse and its climate impact is contingent on the design of the system into which it is integrated. Thus biochar systems are inherently complex and further research is needed to understand their associated opportunities and risks in developing countries. There are a number of reasons why biochar systems .

BioChar

Biochar is the solid product remaining after biomass is heated to temperatures typically between 300°C and 700°C under oxygen-deprived conditions, a process known as "pyrolysis." Conversion of agricultural waste into a powerful soil health enhancer (i.e. biochar) with many promising benefits can be used to conserve cropland diversity, pre- vent deforestation and most importantly to reduce the national food insecurity. Based on various research findings, figure 2 is presenting the benefits of biochar that comprises but not limited to:

1. Increased water retention capacity of soil

2. Reduction in the leaching of water-soluble nutrients

3. Abatement in soil acidity

4. Reduced leaching of nitrogen into ground water

5. Minimized emissions of nitrous oxide

6. Increased cation exchange capacity (CEC) of soil to improve soil fertility

7. Increased persistance of beneficial soil microbes

8. Influencing seed germination, early growth of seedlings and crop production

9. Increased earth-worm abundance, liming effect and priming effect.

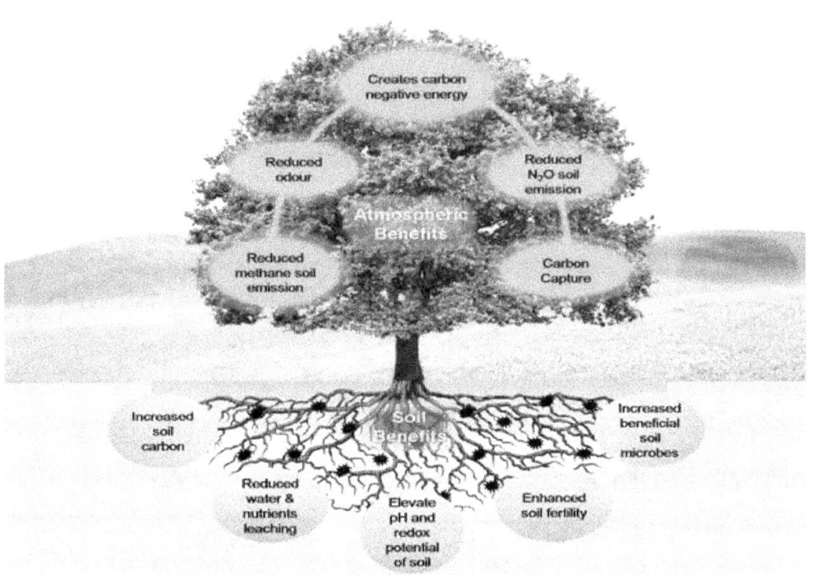

Figure 2: Sustainable application of biochar for atmospheric and soil benefits.

Biochar is a system-defined term referring to black carbon that is produced intentionally to manage carbon for climate change mitigation purposes combined with a downstream application to soils for agricultural effects. It is produced with the intent to be applied to soil as a means of improving soil productivity, carbon storage, or both. Although the term "biochar" has come into common usage only relatively recently, the practice of amending soils with charcoal for fertility management goes back millennia. Instances can be found in Africa, Asia, and notably in the Amazon basin where the historically managed terra preta, or "dark earths," stand out for their capacity to store carbon.

Biochar of maize stalks

Biochar is produced from burning organic material at high temperatures with little to no oxygen availability. The potential of utilizing biochar to sequester carbon in the soil has received considerable research attention in recent years as part of efforts to develop climate smart agricultural practices. As the majority of biochar is carbon (70-80%) it can potentially contribute more carbon than plant residue (approximately 40% carbon) of similar mass. Furthermore, around 60% of this biochar organic carbon is of high stability and therefore resists decomposition more-so than plant material that has not be processed into biochar. That being said, many questions remain as to the effectiveness of biochar application in sequestering carbon. While biochar does contain high levels of carbon, there remains uncertainty as to how long that carbon will persist in the soil following application. The inherent characteristics of the biochar--as dictated by feedstock and pyrolysis conditions--interact with climatic conditions such a precipitation and temperature to influence how long biochar carbon remains stored in the soil.

Recent studies suggest that shorter pyrolysis times and higher pyrolysis temperatures make for more recalcitrant biochar (i.e. it persists for longer periods in the soil). However, there are trade-offs involved as these pyrolysis conditions produce less biochar per unit feedstock. As is so often the case, soil texture plays a key role in determining the persistence of biochar carbon. Biochar becomes stabilized in the soil by interacting with soil particles. Clay particles have more surface area for biochar to interact with and are therefore more effective at stabilizing biochar (Yoni et al., 2016).

Agricultural waste is usually handled as a liability, often because the means to transform it into an asset is lacking. Crop residues in fields can cause considerable crop management problems as they accumulate. The major crop residues produced in India are straws of paddy, wheat, millet, sorghum, pulses (pigeonpea), oilseed crops (castor, mustrad), maize stover and cobs, cotton and jute sticks, sugarcane trash, leaves, fibrous materials, roots, branches and twigs of varying sizes, shapes, forms and densities. Similarly, the agro-industrial residues are rice husk, groundnut shell, cotton waste, coconut shell, coir pith, tamarind shell, mustard husk, coffee husk, cassava peels etc. Some of the common agricultural by-products available in large quantities include bagasse, rice husk, groundnut shell, tea waste, casuarina leaf litter, silk cotton shell, cotton waste, oil palm fibre and shells, cashew nut shell, coconut shell, coir pith etc. (Sugumaran and Sheshadri, 2009).

In India, about 435.98 million tons of agro-residues are produced every year, out of which 313.62 million tons are surplus. These residues are either partially utilized or un-utilized due to various constraints (Murali *et al.*, 2010). Similarly, using different residue to produce ratio values, Koopmans and Koppejan (1997) estimated that about 507,837 thousand tons of field crop residues were generated in India during 1997 of which 43% was rice and 23% wheat.

The estimates from Streets *et al.* (2003) reveal that 16% of total crop residues were burnt. The results from Venkataraman *et al.* (2006) suggest that 116 million tons of crop residues were burnt in India in 2001, but with a strong regional variation. The current availability of biomass in India (2010-2011) is estimated at about 500 million tons/year.

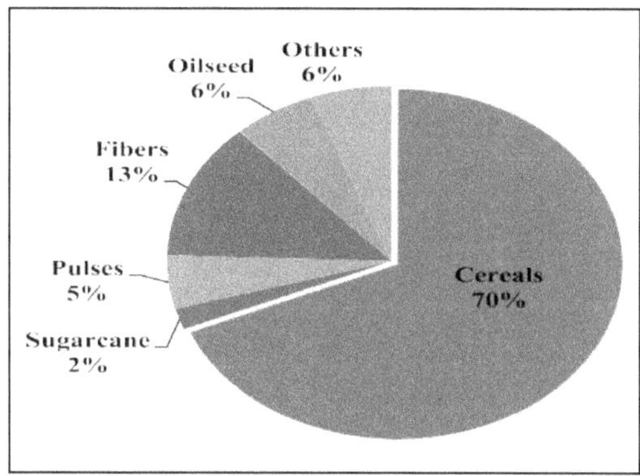

Fig 3. Contribution of various crops to residue generation

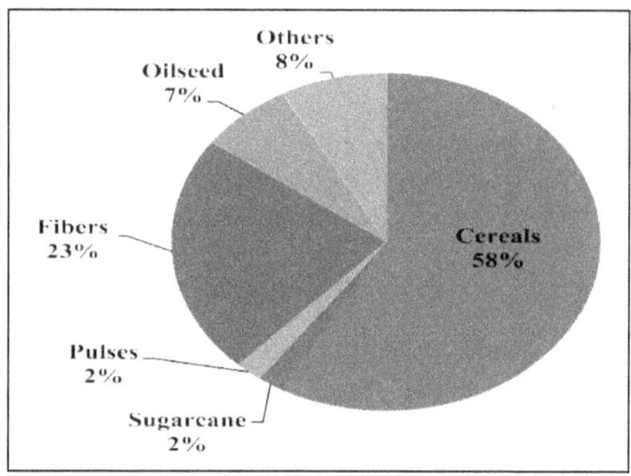

Fig 4. Surplus of various crop residues in India (IARI, 2012) in India (Calculated from MNRE report, 2009)

Among different crops, cereals generate maximum residues (352 Mt), followed by fibres (66 Mt), oilseeds (29 Mt), pulses (13 Mt) and sugarcane (12 Mt). The cereal crops (rice, wheat, maize, millets) contribute 70% while rice crop alone contributes 34% to the crop residues (Fig 3).The surplus residues i.e., total residues generated minus residues used for various purposes, are typically burnt on-farm. Estimated total amount of crop residues surplus in India is 91-141 Mt (IARI, 2012). Cereals and fibre crops contribute 58% and 23%, respectively (Fig 4).

Characteristics of biochar

Characterization of any amendment is the first step to understand the mechanism of action. The properties of biochar are governed by its physical and chemical constituents. The form and size of the feedstock and pyrolysis product may affect the quality and potential uses of biochar (Sohi *et al.*, 2010). The importance of biochar depends on its physical and chemical characteristics, although the relationship of char properties to these applications is not well understood. Using scanning electron microscopy (Figure 5), analysis of morphology of the three biochar products revealed that biomass type influenced morphology under identical pyrolysis conditions Biochar produced from cassava rhizomes and cassava stems have a higher pore density than corncob biochar, while corncob biochar has a high pore size distribution than biochar produced from cassava rhizomes and stems. These results are consistent with measurements of their respective surface areas and total pore volumes. This study revealed that chemical composition of biochar (C, H, O and N), H/C ratio, O/C ratio, the chemical structure (aliphatic and aromatic structures) as well as the product's physical characteristics (surface area, total pore volume and pore size) are associated with biochar's higher cation ion exchange capacity (Saowanee Wijitkosum and Preamsuda Jiwnok ,2019). Biochar's chemical composition, morphology and texture indicated that biochar obtained from biomass feedstocks through slow pyrolysis yielded a biochar product with chemical composition and morphology suitable for soil amendment and carbon sequestration, both directly and indirectly.

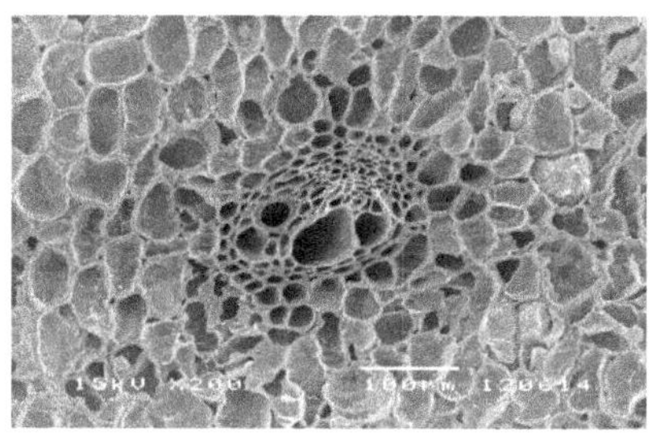

Cornco

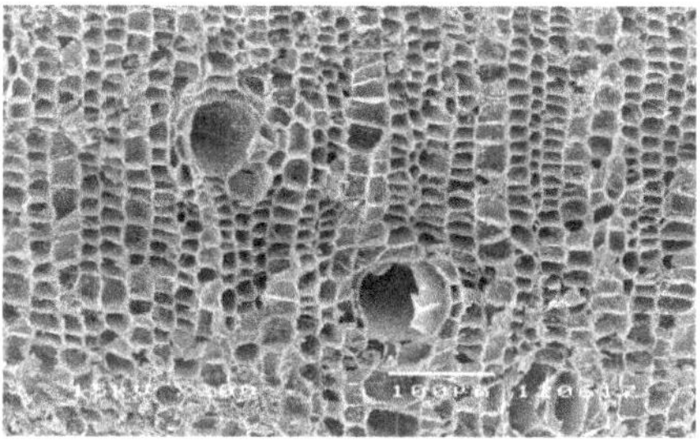

Cassava Rhizome

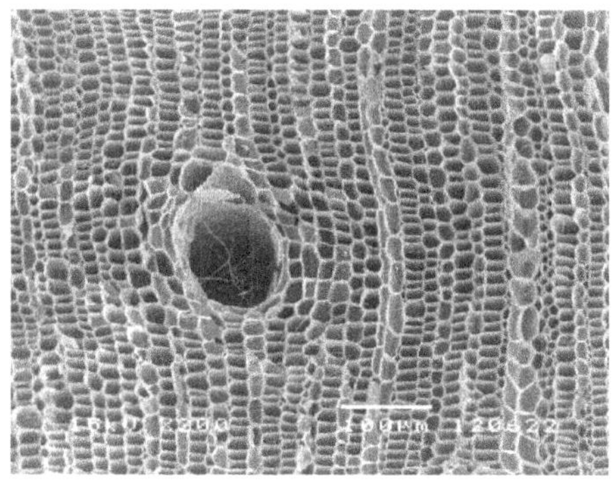

Cassava Stem

Fig-5.Scanning Electron Microscopy (SEM) images showing differences in morphology of biochar produced from different biomass feedstocks

Table -1.Some properties of biochar used in different experiments

Materials used for producing biochar	pH	Total C	Total N	C:N	Ca	Mg	P	K	CEC	Reference
		(%)			(cmol/kg)					
Paper mill waste 1- (waste wood chip)	9.4	50.0	0.48	104	6.2	1.20	-	0.22	9.00	Zwieten et al. (2010)
Paper mill waste 2- (waste wood chip)	8.2	52.0	0.31	168	11.0	2.60	-	1.00	18.00	Zwieten et al. (2010)
Green waste (grass clippings,	9.4	36.0	0.18	200	0.4	0.56	-	21.00	24.00	Chan et al. (2007)

cotton trash, and plant prunings)									
Eucalyptus biochar	-	82.4	0.57	145	-	-	1.87	-	4.69	Noguera et al. (2010)
Cooking biochar	-	72.9	0.76	96	-	-	0.42	-	11.19	Noguera et al. (2010)
Poultry litter (450°C)	9.9	38.0	2.00	19	-	-	37.42	-	-	Chan et al. (2008)
Poultry litter (550°C)	13	33.0	0.85	39	-	-	5.81	-	-	Chan et al. (2008)
Wood biochar	9.2	72.9	0.76	120	0.83	0.20	0.10	1.19	11.90	Major et al. (2010)
Hardwood sawdust	-	66.5	0.3	221	-	-	-	-	-	Spokas et al. (2010)

Adapted from Jha *et al.* (2010)

General chemical properties of biochar samples prepared from different feed stocks are given in Table 1 (Jha *et al.*, 2010). Biochar produced from different feed stock had pH ranged from 8.2-13.0. Invariably, total carbon content of biochar increased with the increase in pyrolysis temperature. Total carbon content in biochar materials produced from different feedstock varied from 33.0 to 82.4%.

Maximum heating temperature and heating rate have a strong influence on the retention of nutrients as does the original composition of the feedstock. N and S compound tends to volatize at a temperature above 200 and 375°C, respectively. So, biochar produced at higher temperature shows depletion of N and S. Whereas, K and P volatilize between 700 and 800°C (DeLuca *et al.*, 2009). High-temperature biochars (800°C) tend to have a higher pH, electrical conductivity (EC), and extractable NO⁻, while low-temperature biochars (350°C) have greater amounts of extractable P, NH4 $^+$, and phenols (DeLuca *et al.*, 2009). The ratios of H/C and O/C decreased with increasing temperature and the lower ratio was found good in terms of aromaticity and stability . There was wide variation in nitrogen content of biochar material produced from different biomass.

Biochar in general had low nitrogen content (0.18-2.0%) and C:N ratio varied from as low as 19 to 221. Except N, the concentration of other elements in biochar increased with the increase in pyrolysis temperature . Biochar contains appreciable quantity of Ca, Mg, K and P. Due to its high pH and appreciable amount of Ca and Mg, biochar acts as liming material (amendment) in acid soils. Lakaria *et al.* (2012) also observed variations in nutrient composition of biochars formed under varying pyrolitic temperature and duration.

BIOCHAR FOR CLIMATE CHANGE MITIGATION

CARBON SEQUESTRATION

Soil C sequestration is the removal of atmospheric CO_2 through photosynthesis to form organic matter, which is ultimately stored in the soil as long-lived, stable forms of C. The global carbon cycle is made up of flows and pools of carbon in the Earth's system. The important pools of carbon are terrestrial, atmospheric, ocean, and geological. The carbon within these pools has varying lifetimes, and flows take place between them all. Carbon in the active carbon pool moves rapidly between pools (Lehmann, 2007b). In order to decrease carbon in the atmosphere, it is necessary to move it into a passive pool containing stable or inert carbon.

Biochar provides a facile flow of carbon from the active pool to the passive pool. In comparison to burning, controlled carbonization converts even larger quantities of biomass organic matter into stable C pools which are assumed to persist in the environment over centuries (Glaser et al. 1998). The conversion of biomass carbon to biochar leads to sequestration of about 50% of the initial carbon compared to the low amounts retained after burning (3%) and biological decomposition (less than 10-20% after 5-10 years) (Lehmann et al. 2006). This efficiency of carbon conversion of biomass to biochar is highly dependent on the type of feedstock, but is not significantly affected by the pyrolysis temperature (within 350-500°C common for pyrolysis). Large amounts of carbon in biochar may be sequestered in the soil for long periods estimated to be hundreds to thousands of years (Lehmann et al. 2006). Among the biochars, maize biochar showed lowest carbon mineralization suggesting its greater potential for long-term carbon sequestration.

Application of biochar showed highest amount of carbon in soil under wheat-pearl millet cropping system. The findings of a recent modeling study (Woolf et al., 2010) reported that biochar amendments to soil, when carried out sustainably, may annually sequester an amount of C equal to 12% the current anthropogenic CO_2 emissions. They estimate that the maximum sustainable technical potential for carbon abatement from biochar is 1-1.8 giga ton (Gt) C per year by 2050. Technical estimates of the potential for biomass pyrolysis coupled with soil storage to sequester carbon suggest that several hundred gigatons of carbon emissions could be sequestered or offset by 2100, which is a large fraction of the total needed to mitigate global climate disruption. It is also easy to monitor carbon sequestration as a climate change mitigation measure for national carbon accounting.

MITIGATION OF GREENHOUSE GAS EMISSIONS

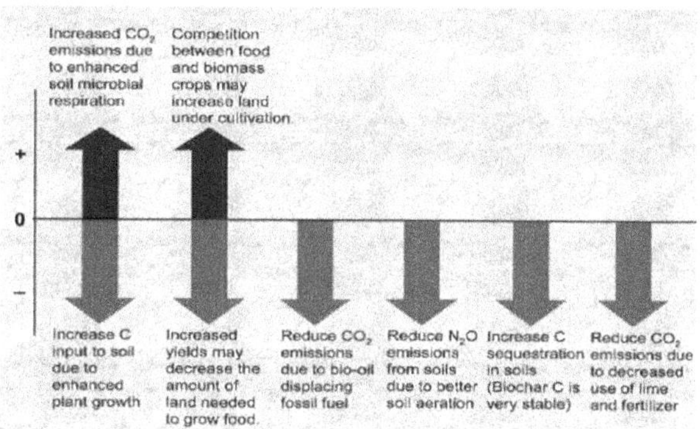

Fig 6. Net impact of biochar applications in soil on greenhouse gas emissions. (Adapted with changes from Rogovska et al., 2008).

Burning of residues emits a significant amount GHGs. For example, 70, 7 and 0.66% of C present in rice straw is emitted as CO_2, CO and CH_4, respectively, while 2.09% of N in straw is emitted as N_2O upon burning. One ton straw on burning releases 3 kg particulate matter, 60 kg CO, 1460 kg CO_2, 199 kg ash and 2 kg SO_2. This change in composition of the atmosphere may have a direct or indirect effect on the radiation balance. Besides other light hydrocarbons, volatile organic compounds (VOCs) and semi-volatile organic compounds (SVOCs) including polycyclic aromatic hydrocarbons (PAHs) and polychlorinated biphenyls (PCBs) and SOx, NOx are also emitted. These gases are important for their global impact and may lead to a regional increase in the levels of aerosols, acid deposition, increase in tropospheric ozone and depletion of the stratospheric ozone layer.

Apart from carbon sequestration, there are other environmental benefits that can be derived from the application of biochar in soils which include reduction in the emission of non-CO_2 GHGs by soils (Fig 6). Soil is a significant source of nitrous oxide (N_2O) and both a source and sink of methane (CH_4). These gases are 23 and 298 times more potent than carbon dioxide (CO_2) as greenhouse gases in the atmosphere. Biochar is reported to reduce N_2O emission could be due to inhibition of either stage of nitrification and/or inhibition of denitrification, or promotion of the reduction of N_2O, and these impacts could occur simultaneously in a soil (DeLuca *et al.*, 2006).

Increased soil aeration from biochar addition reduces denitrification and increases sink capacity for CH_4. Biochar addition induces microbial immobilization of available N in soil, thereby decreasing N_2O source capacity of soil. Increased pH from biochar addition drives N_2 formation from N_2O. When applied to the soil, biochar can lower GHG emissions of cropland soils by substantially reducing the release of N_2O (Lehmann *et al.*, 2003). Reduction of N_2O and CH_4 emission as a result of biochar application is seen to attract considerable attention due to the much higher global warming potentials of these gases compared to CO_2 . Rondon *et al.* (2005) reported a 50% reduction in N_2O emissions from soybean plots and almost complete suppression of CH_4 emissions from biochar amended acidic soils in the Eastern Colombian Plains. Yanai *et al.* (2007), however, reported an 85% reduction in N_2O emission from re-wetted soils containing 10% biochar, compared to soils without biochar. Biochar from municipal biowaste also caused a decrease in emissions of nitrous oxide in laboratory soil chambers (Yanai *et al.* 2007). Spokas *et al.* (2009) also found a significant reduction in N_2O emission in agricultural soils in Minnesota; while Sohi *et al.* (2010) found an emission suppression of only 15%. Additions of 15 g biochar/kg of soil to a grass and 30 g/kg of soil to a soil cropped with soybeans completely suppressed methane emissions (Rondon *et al.* 2005).

An assumption of net carbon abatement of between 0.5 and 1 t CO_2 abatement per ton of feedstock used in pyrolysis biochar system (PBS) as reported by number of published life cycle studies, then this could represent between 35 to 70 million tons of CO_2 abatement per year in India. This assumes however, that all feedstock would be used in PBS, where as in reality there will be many competing uses for the same feedstock and use of biochar might necessitate substitution of residues by a different source of biomass, potentially with attendant greenhouse gas emission that would need to be accounted for. As per the estimates in 2009, the greenhouse gas emission in India was 1,900 million t CO_2 per annum, hence it can be seen that biochar could contribute between 2-4% reduction (Priyadarshini and Prabhune, 2009). More research is needed to understand the interactions between biochar, site specific soil, climatic conditions, and management practices that alter the sink capacity of soils.

INTERACTION OF BIOCHAR WITH SOIL

Interactions with clay minerals Biochar is reported to be found in the organo-mineral fraction of soil, suggesting that biochar interacts with minerals (Liang *et al.*, 2008). Large particles of biochar observed under spectroscope to be embedded within the mineral matrix ,but can also be present as very fine, yet distinguishably particulate, material within aggregates. Rapid association of biochar surfaces with aluminium (Al) and silicon (Si) and, to a lesser extent, with iron (Fe) was found during the first decade after addition of biochar to soil, which increased more slowly within biochar structures (Nguyen et al., 2008). Coating of biochar particles with mineral domains is frequently visible in soils (Lehmann, 2007), and suggests interactions between negatively charged biochar surfaces and either positive charge of variable-charge oxides by ligand exchange and anion exchange, or positive charges of phyllosilicates by cation bridging. Likewise calcium (Ca) can increase the biochar stability, most likely by enhancing interactions with mineral surfaces . Large amounts of ionic iron (Fe) and aluminium (Al) were also found in biochar type humic fractions (Nakamura et al., 2007), which may indicate that complexion between biochar surfaces and polyvalent metal ions could increase biochar stability. The energy dispersive X-ray spectroscopy (EDS) analysis indicated that pearl millet stalk biochar and rice straw biochar particles consisted of high calcium agglomerates (Purakayastha et al., 2015).

Biochar stability The stability of biochar is of prime requisite for long-term sequestration of C in soil. The mechanisms of biochar stability is mainly due to the composition changes through a complete destruction of cellulose and lignin, thus changes the appearance of aromatic structures (Paris et al., 2005) with furan-like compounds (Baldock and Smernik, 2002). These changes in the composition of organic bonds by pyrolysis have a significant effect on the stability of biochar. The conversion of organic matter to biochar by pyrolysis significantly increases the recalcitrance of C in the biomass. The principal mechanisms operating in soils through which biochar entering the soil is stabilized and significantly increase its residence time in soil are intrinsic recalcitrance, spatial separation of decomposers and substrate, and formation of interactions between mineral surfaces (Sollins et al., 1996). Among the four different biochar used for CO_2 efflux study, the maize biochar was found to be the most stable showing reduced C mineralization to protect the native soil organic C (Purakayastha et al., 2015). The reduced C mineralization was also observed in the case of pearl millet and wheat biochar. Contrarily, rice biochar exhibited higher C mineralization.

CARBON SEQUESTRATION AND GREENHOUSE GAS EMISSION

Biochar is carbon negative and thus resulting in long-term removal of carbon from the atmosphere. Mitigation of carbon emissions is obtained not only from biochar soil application, but also from substitution of fossil fuel by the produced bio-oil. As discussed, the stability biochar could be increased by raising the pyrolysis temperature, but this will be at the expense of the quantity produced. This inverse relation makes it possible to determine the pyrolysis temperature that gives the highest carbon sequestration.

Interestingly, the highest biochar carbon sequestration is achieved at 500 °C, despite the fact that biochar made at higher temperatures is relatively more recalcitrant than low temperature biochars. It has been projected that in India about 309 million tons of biochar (eqv. to 154 mt of biochar C) could be produced annually, the application of which might offset about 50% of C emission (292 Tg C yr-1) from fossil fuel (Lal, 2005). Additionally both heat and gases can be captured during production of biochar by pyrolysis to produce energy carriers such as electricity, bio-oil, or hydrogen and certain other valuable co-products. The potential of biochar application for soil organic carbon (SOC) sequestration may be 1 Pg C yr-1 (Sohi et al., 2010) or more (Lehmann et al., 2006). Biochar can have $\geq$60-80% carbon composition that is equivalent to $\geq$2.20-2.94 t carbon dioxide sequestered ton-1 biochar (Verma et al., 2014).

CONCLUSIONS

Atmospheric concentration of CO_2 at 390 ppm in 2010, and increase at the rate of 2.3 ppm/yr is exacerbating the risks of global warming, soil degradation and desertification and environmental pollution. C sequestration in soils and vegetation, with a technical potential of ~3 Pg C/yr with a drawdown of atmospheric CO_2 by 50 ppm by the end of the twenty-first century, is a cost-effective option with numerous ancillary or co-benefits especially through enhancement of ecosystem services. Important among the co-benefits of enhancing the SOC pool are improved soil quality and the attendant increase in agronomic production . C sequestration enhances soil quality and the associated water and nutrient cycles and thereby it enhances the productive potential of the land – on which all terrestrial life depends. Biochar production and application to soil has potential advantages as soil amendment covering benefits beyond carbon sequestration. This includes improvement of soil physical properties that benefit crops, improved retention and availability of soil nutrients, improved biological activity and consequently higher crop yields and societal advantages through mitigation of global warming through carbon sequestration

References

➤ DeLuca, T.H., MacKenzie, M.D., Gundale, M.J. and Holben,W.E. 2006. Wildfire-produced charcoal directly influences nitrogen cycling in forest ecosystems. *Soil Science Society America Journal*, 70: 448-453.

➤ Glaser, B., Haumaier, L., Guggenberger, G. and Zech, W. 1998. Black carbon in soils: The use of benzene carboxylic acids and specific markers. *Organic Geochemistry*, 29: 811-819.

➤ Koopmans, A. and Koppejan, J. 1997. Agricultural and forest residues – generation, utilisation and availability. Regional Consultation on Modern Applications of Biomass Energy. 6.

➤ Lakaria, B.L., Sumitra Giri, Pramod Jha and Biswas, A.K. 2012. Biochar preparation and characterization. (*Unpublished*).

➤ Lal, R., 2005. Carbon sequestration and climate change with specific reference to India. Proc. Int. Conf. on Soil, Water and Environmental Quality-Issues and Strategies. Indian Soc. Soil Sci., Division of SS and AC, IARI, New Delhi, India, pp. 295-302.

➤ Lehmann, J. and Rondon, M. 2006. Biochar soil management on highly weathered soils in   the humid tropics. In: *Biological*

Approaches to Sustainable Soil Systems (N. Uphoff *et al.* eds), Boca Raton, FL: CRC Press. pp 517-530.

➢ Lehmann, J., Pereira da Silva Jr. J., Steiner, C., Nehls, T., Zech, W. and Glaser, B. 2003. Nutrient availability and leaching in an archaeological Anthrosol and a Ferralsol of the Central Amazon basin: fertilizer, manure and charcoal amendments. *Plant and Soil*, 249: 343- 357.

➢ Liang, B., Lehmann, J., Solomon, D., Sohi, S., Thies, J.E., Skjemstad, J.O., Luizão, F.J., Engelhard, M.H., Neves, E.G., Wirick, S., 2008. Stability of biomass-derived black carbon in soils. Geochmica et Cosmochimica Acta 72, 6078-6096.

➢ Murali, S., Shrivastava, R. and Saxena, M. 2010. Greenhouse gas emissions from open field burning of agricultural residues in India. *Journal of Environmental Science and Engineering*, 52(4): 277-84.

➢ Nguyen, B., Lehmann, J., Kinyangi, J., Smernik, R., Engelhard, M.H., 2008. Long-term black carbon dynamics in cultivated soil. Biogeochemistry 89, 295-308.

➢ Paris, O., Zollfrank, C., Zickler, G.A., 2005. Decomposition and carbonisation of wood biopolymers -a microstructural study of softwood pyrolysis. Carbon 43, 53-66.

➢ Priyadarshani, K. and Prabhune, R. 2009, Biochar for carbon reduction, sustainable agriculture and soil management (BiocharM), final report for APN Project, ARCP, 2009-12 NSY-Karve.

➢ Rondon, M., Ramirez, J.A. and Lehmann, J. 2005. Charcoal additions reduce net emissions of greenhouse gases to the atmosphere. In: *Proceedings of the 3rd USDA Symposium on*

Greenhouse Gases and Carbon Sequestration, Baltimore, USA, March 21-24, 2005.

➢ Saowanee Wijitkosum and Preamsuda Jiwnok.2019. Elemental Composition of Biochar Obtained from Agricultural Waste for Soil Amendment and Carbon Sequestration.

➢ Sohi, S., Krull, E., Lopez-Capel, E. and Bol, R. 2010. A review of biochar and its use and function in soil. *Advances in Agronomy*, 105: 47-82.

➢ Sollins, P., Homann, P., Caldwell, B.A., 1996. Stabilization and destabilization of soil organic matter mechanisms and controls. Geoderma 74, 65-105.

➢ Spokas, K.A., Koskinen, W.C., Baker, J.M. and Reicosk, D.C. 2009. Impacts of woodchip biochar additions on greenhouse gas production and sorption/degradation of two herbicides in a Minnesota soil.   *Chemosphere*, 77: 574-581.

➢ Sugumaran, P. and Sheshadri, S. 2009. Evaluation of selected biomass for charcoal production. *Journal of Scientific & Industrial Research*, 68: 719-723.

➢ Verma, M., M'hamdi, N., Dkhili, Z., Brar, K.S., Misra, K., 2014. Thermo-chemical transformation of agro-biomass into biochar simultaneous carbon sequestration and soil amendment, biotransformation of waste biomass into high value. Biochemicals 51-70.

➢ Woolf, D., Amonette, J.E., Street-Perrott, F.A., Lehmann, J and Joseph, S. 2010. Sustainable biochar to mitigate global climate change. *Nature Communications*, 1: 1-9.

➢ Yanai, Y., Toyota, K. and Okazani, M. 2007. Effects of

charcoal addition on N_2O emissions from soil resulting from rewetting air-dried soil in short-term laboratory experiments. *Soil Science and Plant Nutrition*, 53: 181-188.

➢ Yang, F., et al. (2016). "The Interfacial Behavior between Biochar and Soil Minerals and Its Effect on Biochar Stability." Environmental Science & Technology 50(5): 2264-2271.

Chapter 6

SEQUESTERING CARBON IN FOREST, GRASSLAND, AND WETLAND ECOSYSTEM IS THE WAY TO MITIGATE CLIMATE CHANGE

R.Murugaragavan[1], S.S.Rakesh[2], S.R.Shrirangasami[3], R.Elangovan[4], B.Balamurali[5], P.T.Ramesh[6] and S.Saravanakumar[7]

1 Teaching Assistant in Environmental Science, Department of Soils and Environment, AC&RI, Tamil Nadu Agricultural University, Madurai-625104

2 Doctoral Research Scholar in Environmental Science, Department of Environmental Sciences, TNAU, Coimbatore-641003

3Assistant Professor in Agronomy, Rice Research Station, TNAU, Ambasamudram-627401

4Assistant Professor, College of Agricultural Technology, TNAU, Theni-625562, Tamil Nadu, India

5 Subject Matter Specialist in Agricultural Meteorology, Krishi Vigyan Kendra, TNAU, Papparapatty-636809, Dharmapuri District

6 Associate Professor in Environmental Science, Agricultural College and Research Institute, TNAU, Killikulam-628252

7 Project Scientist, National Remote Sensing Centre, Hyderabad, Telungana-500037

INTRODUCTION

The major terrestrial ecosystem, forests occupy 30–43% of the world's land surface. Forests are an important source for fiber and fuel, for human consumption, and of vital importance for the economy of many regions in the world. Forests are important to our quality of life and society, as they provide many resources and services including mitigation of climate change. Trees take up CO_2 in the atmosphere and release oxygen (O_2) through photosynthesis. The sum of photosynthesis over a year is termed gross primary productivity (GPP). The fixed carbon is then transferred to their stems, roots, and leaves for growth. The actual carbon fixed into plants, the net primary productivity (NPP) of an ecosystem, is the balance between GPP and the carbon lost through plant respiration (i.e., the construction and maintenance cost). When leaves and branches fall and decompose, the stored carbon will be released back to atmosphere through respiration. Part of dead organic matter will be transferred to the soil. When carbon losses by microbial respiration (heterotrophic respiration, redox potential) in litter and soil are accounted for obtaining the net carbon balance of an ecosystem. This net balance is termed as net ecosystem productivity (NEP). Forest ecosystems play a major role in the global carbon cycle by storing large amount of carbon in live plant biomass, dead plant material, and the soils and are considered as carbon sinks.

Grasslands, including rangelands, shrublands, pastureland, and cropland sown with pasture and fodder crops, covered approximately 3.5 billion ha in 2000, representing 26 percent of the world land area and 70 percent of the world agricultural area, and containing about 20 percent of the world's soil carbon stocks (FAOSTAT, 2009; Ramankutty *et al.*, 2008; Schlesinger, 1977). People rely heavily upon grasslands for food and forage production. Around 20 percent of the world's native grasslands have been converted to cultivated crops (Ramankutty et al., 2008) and significant portions of world milk (27 percent) and beef (23 percent) production occur on grasslands managed solely for those purposes. The livestock industry largely based on grasslands and it provides livelihoods for about 1 billion of the world's poorest people and one-third of global protein intake (FAO, 2006).

Carbon Sequestration in Forest Ecosystem:

Forest contributes lot of additional benefits such as habitats for wildlife, clean water and carbon storage, and climate mitigation. Forest biomass is the major pool of vegetation carbon. The total amount of carbon sequestered in forest vegetation is approximately 359 billion tons (Allen *et al.,* 2010). Forest soil is another carbon pool that stores large amount of carbon. Overall, the amount of carbon stored in the forest ecosystems is twice that in the atmosphere (Lal, 2005). As forests play an important role in the global carbon cycling and curbing climate change. Due to fossil fuel burning and deforestation, concentration of atmospheric carbon dioxide (CO_2) has increased from 280 ppm at preindustrial revolution to 394 ppm presently. As a result, average global surface temperature is increasing at an unprecedented rate. These changes as well as changes in precipitation, nitrogen deposition are likely to have significant effects on tree growth and forest dynamics, and eventually, carbon accumulation in forest ecosystems (Zuidema *et al.,* 2013).

Agroforestry and Carbon sequestration

Agroforestry enhances carbon uptake by lengthening the growing season, expanding the niches from which water and soil nutrients are drawn and, in the case of nitrogen (N)-fixing species, enhancing soil fertility. The result is that when agroforestry systems are introduced in suitable locations, carbon is sequestered in the tree biomass and tends to be sequestered in the soil as well (Jose, 2009). Improved management in existing agroforestry systems could sequester 0.012 Tg C yr^{-1} while conversion of 630 million ha of unproductive or degraded croplands and grasslands to agroforestry could sequester as much as 0.59 Tg C annually by 2040 (IPCC, 2000), which would be accompanied by modest increases in N_2O emissions as more N circulates in the system. Using seeded grasses for cover cropping, catch crops and more complex crop rotations all increase carbon inputs to the soil by extending the time over which plants are fixing atmospheric CO_2 in cropland systems. Rotations with grass, hay or pasture tend to have the largest impact on soil carbon stocks (West and Post, 2002). Adding manure to soil builds soil organic matter in grasslands. The synthesis by Smith et al. (2008) suggests that adding manure or biosolids to soil could sequester between 0.42 and 0.76 t C ha^{-1} yr^{-1} depending on the region (sequestration rates tend to be greater in moist regions than in dry).

Grass land ecosystem and carbon sequestration

Removing biomass, changing the vegetation or altering soil function is an integral part of traditional grassland management systems, which fosters dependable yields of forage. However, disturbance through overgrazing, fire, invasive species, etc. can also deplete grassland systems of carbon stocks (Smith et al., 2008). Harvesting a large proportion of plant biomass enhances yields of useful material (e.g. for forage or fuel), but decreases carbon inputs to the soil .

Primary production in overgrazed grasslands can decrease if herbivory reduces plant growth or regeneration capacity, vegetation density and community biomass, or if community composition changes. If carbon inputs to the soil in these systems decrease because of decreased net primary production or direct carbon removal by livestock, soil carbon stocks will decline. Like carbon sequestration in forests or agricultural land, sequestration in grassland systems primarily, but not entirely in the soils is brought about by increasing carbon inputs. It is widely accepted that continuous excessive grazing is detrimental to plant communities and soil carbon stocks. When management practices that deplete soil carbon stocks are reversed, grassland ecosystem carbon stocks can be rebuilt, sequestering atmospheric CO_2.

In grasslands, carbon assimilation is directed towards the production of fibre and forage by manipulating species composition and growing conditions. Ecosystems are a major source and sink for the three main biogenic greenhouse gases (GHG) CO_2, nitrous oxide (N_2O) and methane (CH_4). In undisturbed ecosystems, the carbon balance tends to be positive, carbon uptake through photosynthesis exceeds losses from respiration, even in mature, old growth forest ecosystems. Disturbance, such as fire, drought, disease or excessive forage consumption by grazing, can lead to substantial losses of carbon from both soils and vegetation (Adams *et al.*, 2009). Disturbance is a defining element of all ecosystems that continues to influence the carbon uptake and losses that determine long term ecosystem carbon balance (Randerson *et al.*, 2002). Human land use activities function much like natural activities in their influence on ecosystem carbon balance. CO_2 is produced when forest biomass is burned, and soil carbon stocks begin to decline soon after soil disturbances (Lal, Kimble and Stewart, 2000). Like natural disturbances such as fire and drought, land use change affects vegetation and soil dynamics, often prompting further increased carbon releases and decreased carbon uptake. Deforestation, degradation of native grasslands and conversion to cropland have prompted losses of biomass and soil carbon of 450–800 Gt/CO_2 equivalent to 30–40 percent of cumulative fossil fuel emissions (Hickler, 2008) Emissions from conversion from forests to cropland or other land use have dominated carbon losses from terrestrial ecosystems (DeFries *et al.,* 1999), but substantial amounts of carbon have been lost from biomass and soils of grassland systems as well (Shevliakova *et al.,* 2009). The basic processes governing the carbon balance of grasslands are similar to those of other ecosystems: the photosynthetic uptake and assimilation of CO_2 into organic compounds and the release of gaseous carbon through respiration (primarily CO_2 but also CH_4). Biomass in grassland systems, being predominantly herbaceous (i.e. non woody), is a small, transient carbon pool (compared to forest) and hence soils constitute the dominant carbon stock. Grassland systems can be productive ecosystems, but restricted growing season length, drought periods and grazing-induced shifts in species composition or production can reduce carbon uptake relative to that in other ecosystems. Soil organic carbon stocks in grasslands have been depleted to a lesser degree than for cropland and in some regions biomass has increased due to suppression of disturbance and subsequent woody encroachment. Much of the carbon lost from agricultural land soil and biomass pools can be recovered with changes in management practices that increase carbon inputs, stabilize carbon within the system or reduce carbon losses, while still maintaining outputs of fibre and forage.

Grassland management to enhance production (through sowing improved species, irrigation or fertilization), minimizing the negative impacts of grazing or rehabilitating degraded lands can each lead to carbon sequestration. Improved grazing management

(management that increases production) leads to an increase of soil carbon stocks by an average of 0.35 Mg C ha^{-1} yr^{-1} (Conant, Paustian and Elliott, 2001).

An important argument in favour of grassland carbon sequestration is that implementation of practices to sequester carbon often lead to increased production and greater economic returns. Forage removal practices that disturb the system and prompt carbon losses usually reflect attempts to enhance forage utilization, but the complement is not necessarily true: practices that sequester carbon do not necessarily result in reduced forage utilization. Reducing the amount of carbon inputs removed, or increasing production, carbon inputs or below-ground allocation, could all lead to increasing soil carbon stocks. Grazing management can lead to decreased carbon removal if grazing intensities are reduced or if grazing is deferred while forage species are most actively growing. Sustainable grazing management can thus increase carbon inputs and carbon stocks without necessarily reducing forage production. Grazing management can also be used to restore productive forage species, further augmenting carbon inputs and soil carbon stocks. Other practices that enhance production, such as sowing more productive species or supplying adequate moisture and nutrients, also result in greater carbon uptake, ecosystem carbon stocks and forage production.

Climate Mitigation measures in grasslands

Grasslands contain a substantial amount of the world's soil organic carbon. Integrating data on grassland areas and grassland soil carbon stocks results in a global estimate of about 343 billion tonnes of C nearly 50 percent more than is stored in forests worldwide (FAO, 2007). Just as in the case of forest biomass carbon stocks, grassland soil carbon stocks are susceptible to loss upon conversion to other land uses or following activities that lead to grassland degradation (e.g. overgrazing). Current rates of carbon loss from grassland systems are not well quantified. Over the last decade, the grassland area has been diminishing while arable land area has been growing, suggesting continued conversion of grassland to croplands (FAOSTAT, 2009). When grasslands are converted to agricultural land, soil carbon stocks tend to decline by an average of about 60 percent. Grassland degradation has also expanded, probably contributing to the loss of grassland ecosystem carbon stocks. Arresting grassland conversion and degradation would preserve grassland soil carbon stocks. The magnitude of the impact on atmospheric CO_2 is much smaller than that due to deforestation, but preserving grassland soil carbon stocks serves to maintain the productive capacity of these ecosystems that make a substantial contribution to livelihoods.

Grass land as carbon stock

1. Grazing management can be improved to reverse grazing practices that continually remove a very large proportion of aboveground biomass. Implementing a grazing management system that maximizes production, rather than offtake, can increase carbon inputs and sequester carbon.

2. Sowing improved species can lead to increased production through species that are better adapted to local climate, more resilient to grazing, more resistant to drought and able to enhance soil fertility (i.e. N-fixing crops). Enhancing production leads to greater carbon inputs and carbon sequestration.

3. Direct inputs of water, fertilizer or organic matter can enhance water and N balances, increasing plant productivity and carbon inputs,

potentially sequestering carbon. Inputs of water, N and organic matter all tend to require energy and can each enhance fluxes of N2O, which are likely to offset carbon sequestration gains.

4. Restoring degraded lands enhances production in areas with low productivity, increasing carbon inputs and sequestering carbon.

5. Including grass in the rotation cycle on arable lands can increase production return organic matter (when grazed as a forage crop), and reduce disturbance to the soil through tillage. Thus, integrating grasses into crop rotations can enhance carbon inputs and reduce decomposition losses of carbon, each of which leads to carbon sequestration.

Improved management techniques can increase forage production and reduce feed costs, financially benefiting producers. As forage production increases, an ancillary benefit may lie in increased sequestration of atmospheric carbon. Indeed, Gifford et al. (1992) noted that improved pasture management is an important consideration when computing a national carbon budget. A variety of grassland management practices lead to near term increases in both production and sequestration of carbon, and practices that sequester carbon often enhance producer income. Practices that reduce offtake, through grazing or harvest is tend to enhance carbon inputs, building carbon stocks. Thus, grazing management practices that increase carbon inputs by increasing production can sequester carbon. Also, practices that increase production inputs by enhancing soil fertility or sowing more productive species can help to build up soil carbon stocks. Directly introducing more carbon to the system through organic matter (e.g. manure) additions will also lead to increased carbon stocks, although it has been pointed out that increases are gained at the expense of carbon inputs where feed crops are grown.

In addition to enhancing forage production and food security, many land management practices that sequester carbon prompt other changes in environmental processes that are beneficial for other reasons. Practices that sequester carbon in grassland soils tend to maximize vegetative cover, reducing wind and water-induced erosion. Reducing sediment load increases water quality while reducing airborne particulate matter improves air quality. Carbon sequestering practices can also enhance ecosystem water balance; building soil organic matter stocks tends to enhance water infiltration and soil moisture status in arid, semi-arid environments. In many cases practices that sequester carbon can lead to greater biodiversity.

Most grassland management practices with the potential to sequester carbon were developed to address issues other than carbon sequestration. For example, expanding grasslands through agricultural set-asides and rehabilitating degraded rangelands are often intended to arrest wind and water erosion. Practices that preserve the habitat, like grassland preservation, rehabilitation, etc., can preserve species and biodiversity. A variety of practices that integrate grass species into arable crop rotation (for example, catch crops used to retain nutrients, cover crops to reduce erosion, grass crops in rotation) sequester carbon and also retain nutrients in agricultural systems, reducing downstream pollution (Stevens and Quinton, 2009).

Carbon sequestration in Silvipasture

In general, establishment of forests on cropland, heathland or degraded land results in significant increases in soil organic carbon, whereas conversion of grassland to forest is less effective. Secondary forest development after logging of primary forest can restore organic soil stocks, but conversion of native forest to plantation often leads to losses of soil carbon.

Sequestration of carbon in forest soil is highly dependent on site properties, land use history and tree species used. Mixed stands can be superior to pure stands in this respect, and less intensive management, as well as silvopastoral management systems, can lead to increased sequestration of organic carbon. Agroforestry systems in the tropics also have a positive impact on soil carbon sequestration. Increased temperatures due to global warming can lead to increased carbon sequestration in forest soil as a result of higher productivity, but this may be counteracted by increased losses of organic carbon by decomposition due to higher soil temperatures and moisture.

The required informations are needed on several factors in order to optimize carbon sequestration in forest soil. This includes the effects of climate change, tree species diversity, forest management system, soil properties, and belowground processes and the role of soil biota. We need also to better understand how we can direct organic matter into stable mineral soil carbon pools.The forest ecosystem stores huge quantities of carbon in biomass, which is important to offset net carbon emission. Forest biomass can also be used to replace fossil fuels. Therefore, it is important to account for the whole carbon budget of the forest ecosystem.

Carbon sequestration and Wetlands

Wetlands cover about 3% of the global land area, but contain 20–30% of the terrestrial stocks of soil organic carbon. It is highly important to protect these vulnerable stocks which are seriously threatened by drainage and climate change. Wetland protection and restoration (rewetting) needs to be a significant factor in climate policy and now UNFCCC has accepted to include wetland restoration as valid action to reduce net GHG emission.In the landscape, organic soil wetlands are an integrated part of a network of forests and croplands on higher or drier sites and often linked by streams and lakes. A considerable part of the soil organic carbon can be transported by water, mostly by horizontal transport within surface soil layers.

Improving Carbon Sequestration by Terrestrial Ecosystems

1. The study suggests that Joint Forest Management in India could be effectively utilized for carbon sequestration so as to mitigate climate change. Forestry can play a major role towards increasing the global carbon sequestration if the world's forest could be managed properly with due importance to afforestation and reforestation and carbon management in existing forests.

2. The carbon assimilation is altered by environmental disturbances like wild fires and hurricanes by decreasing carbon uptake or increasing carbon release into the atmosphere.

3. Agroforestry interventions, because of their ability to provide economic and environmental benefits, are considered to be the best "no regrets" measures in making communities adapt and become resilient to the impacts of climate change. Agroforestry practices like alley cropping and silvopastures have the greatest potential for conserving and sequestering carbon because of the close interaction between crops, pasture, trees and soil. The important elements of agroforestry systems that can play a significant role in the adaptation to climate change include changes in the microclimate,

protection through provision of permanent cover, opportunities for diversification of the agricultural systems, improving efficiency of use of soil, water and climatic resources, contribution to soil fertility improvement, reducing carbon emissions and increasing sequestration.

4. The International Panel on Climate Change (IPCC) estimates that the current worldwide area under agroforestry is 400 million ha, which results in a carbon gain of 0.72 $Mgha^{-1}$ $year^{-1}$. It is estimated that the potential carbon gain could increase to 26×10^6 $Mgha^{-1}$ $year^{-1}$ by 2010 and to 45×10^6 $Mgha^{-1}$ $year^{-1}$ by 2040.

5. The use of agroforestry crops is a promising tool for reducing atmospheric CO_2 concentration through fossil fuel substitution. In particular, plantations characterized by high yields such as short rotation forestry are becoming popular worldwide for biomass production and their role acknowledged in the Kyoto Protocol

6. A study of carbon storage and nitrogen cycling in silvopastoral systems on sodic soils. They observed that compared to 'grass-only' systems, soil organic matter, biological productivity and carbon storage were greater in the silvopastoral systems. Of the total nitrogen uptake by the plants, 4 to 21 per cent was retained in the perennial tree components and nitrogen cycling in the soil-plant system was found to be efficient. Thus, they suggested that the silvopastoral systems, integrating trees and grasses, hold promise as a strategy for improving highly sodic soils.

7. Urban forests are very important as far as carbon sequestration is concerned. They have a great potential of carbon

sequestration which can be further increased by their proper management.

8. Bamboo forests play an important role as far as carbon sequestration is concerned. The bamboo forests are also included in the list of eligible afforestation and reforestation projects under the Clean Development Mechanism.

9. As a major non wood forest product and wood substitute, bamboo is of increasing interest to ecologists owing to its rapid growth and correspondingly high potential for mitigating climate change. Compared with other types of forests, the bamboo forests generate different ecosystem services, such as carbon storage, and water and soil conservation because of their special root reporting regeneration strategy and selective cutting utilization system. Besides, bamboo forests also aid in cleaning air, reducing noise pollution and maintaining wildlife biodiversity. Their carbon sequestration potential and other services can be enhanced by their proper management. Land areas set aside to be preserved in their natural habitat and designated as national parks play a significant role in sequestering carbon.

Climate change mitigation

Mitigation investments are crucially important for reducing the impacts of climate change, but GHG concentrations will continue to increase for decades despite implementation of even the most aggressive climate policies. Because yield reductions under drought, heat stress, floods and other extreme events will be the most consequential, negative impacts of climate change, efforts to adapt to a changing climate should focus on increasing the resilience of management systems. The increasing frequency of droughts in the drylands and droughts of longer duration are expected to have a substantial negative effect on the sustainability and viability of livestock production systems in semiarid regions. Grassland management practices maximize the infiltration, capture and utilization of precipitation for production. In cases where sustainable grazing management increases soil carbon stocks, soil water holding capacity increases. Both facets of enhancing water balance will increase drought resilience.

Grassland management practices that sequester carbon tend to make systems more resilient to climate variation and climate change: increased soil organic matter (and carbon stocks) increases yields. Soil organic matter also enhances soil fertility, reducing reliance on external N inputs. Surface cover, mulch and soil organic matter all contribute to a decrease in inter annual variation in yields (Lal *et al.*, 2007) and practices that diversify cropping systems, such as grass and forage crops in rotation, sequester carbon and enhance yield consistency.

Agricultural practices intended to mitigate GHG emissions could increase vulnerability to climate variation and climate change, if they increase the energy supply from food production systems (e.g. to supply biomass energy), or prevent arable land from being cultivated (e.g. afforestation).

However, practices that minimize soil disturbance and maintain good ground cover, restore soil carbon stocks and related soil biological activity, diversify crops and integrate crop/livestock production, will tend to increase soil carbon stocks and enhance resilience to drought and climate change (Woodfine, 2009).

Carbon sequestration and income

The Intergovernmental Panel on Climate Change (IPCC) (2007b) estimated that grasslands, forestry and agriculture would sequester approximately 8 Gt CO_2 yr^{-1} given carbon prices of USD100/Mg CO_2, including reduced emissions from deforestation and degradation would maintain an additional 4 Gt CO_2 yr^{-1} in the soil, raising total contribution of the land sectors to about one third of total annual global emissions (i.e. 12 Gt CO_2 yr^{-1} out of 30 Gt CO_2 yr^{-1}). Substantial amounts of CO_2 emission from the land sector and large potential for sequestration with changes in land management are among the most important arguments in favour of terrestrial sequestration. Some practices that sequester carbon require land managers to forego optimal harvest (e.g. reducing forage offtake), tolerate reduced yields (e.g. reduced stocking rates) or change land use (e.g. cessation of grazing of vulnerable soils). Others require investments in new equipment that could be substantial (e.g. for seeding, irrigation or fertilization).

However, the primary investments necessary for successful widespread adoption of many of the land management practices that enhance ecosystem carbon storage are knowledge, education and information. Most of the materials required for the implementation of many practices that sequester carbon (e.g. improved species, legumes, grazing management, fire management, etc.) are often no different than those required for degradative land management practices – they differ primarily in their implementation. Carbon emissions from land-use change arise primarily from countries that are exempt from emission reductions under the Kyoto Protocol. Widespread disturbance and degradation and continuing deforestation make carbon sequestration and preservation (i.e. United Nations Collaborative Programme on Reducing Emissions from Deforestation and Forest Degradation [UN-REDD]) substantial sequestration opportunities in these developing countries (Benitez et al., 2007). Engagement of developing countries in emission reduction activities that simultaneously enhance adaptive strategies is another argument in favour of grassland carbon sequestration (Jung, 2005). Given modest costs and the use of existing technologies, grassland carbon sequestration in developing countries could be enacted in the near term, offsetting emissions from other sectors now, allowing time for the larger investments required to reduce directly emissions from burning fossil fuels.

Investments in carbon sequestering practices in developing countries that increase grassland or livestock efficiency or productivity and reduce vulnerability to impacts of climate change (i.e. enhancing adaptation) are likely to promote relatively immediate sustainable returns. The economic, environmental and social costs of land degradation are substantial and investments in sustainable grassland management tend to be an efficient use of limited development resources.

Conclusion

The challenges of climate change can be efficiently reduced by the storage of carbon in forest, grassland, wetland and agroforestry ecosystem for longer periods of time. Adoption of carbon sequestration measures can considerably reduce the increase in atmospheric CO_2 level. In order to sustain the amount of carbon in the forest, grassland, wetland and agroforestry ecosystem through different conservation measures viz., good land and forest management practices. This is the best way to mitigate climate change in the world.

References

Adams, H.D., Guardiola-Claramonte, M., Barron-Gafford, G.A., Villegas, J.C., Breshears, D.D., Zou, C.B., Troch, P.A. & Huxman, T.E.2009. Temperature sensitivity of drought-induced tree mortality portends increased regional die-off under global-change-type drought. Proc. Natl. Acad. Sci. USA,106: 7063–7066.

Allen CD, Macalady AK, Chenchouni H et al (2010) A global overview of drought and heatinduced tree mortality reveals emerging climate change risks for forests. For Ecol Manag.259:660–684

Benitez, P.C., McCallum, I., Obersteiner, M. & Yamagata, Y.2007. Global potential for carbon sequestration: geographical distribution, country risk and policy implications. Ecological Economics,60: 572–583

Conant, R.T., Paustian, K. & Elliott, E.T. 2001. Grassland management and conversion into grassland: effects on soil carbon. Ecol. Appl.,11: 343–355.

DeFries, R.S., Field, C.B., Fung, I., Collatz, G.J. & Bounoua, L.1999. Combining satellite data and biogeochemical models to estimate global effects of human-induced land cover change on carbon emissions and primary productivity. Global Biogeochem. Cycles,13: 803–815

FAO. 2006. Livestock's long shadow: environmental issues and options. Rome.

FAO.2007. State of the World's Forests2007. Rome.

FAOSTAT.2009. Statistical Database 2007. Rome.

Gifford, R.M., Cheney, N.P., Noble, J.C.R.J.S., Wellington, A.B. & Zammit, C.1992. Australia's renewable resources: sustainability and global change.In: R.M. Gifford & M.M. Barson, eds. pp. 151–188. Australian Government Publishing Service.

Hickler T, Smith B, Prentice IC et al (2008) CO_2 fertilization in temperate FACE experiments not representative of boreal and tropical forests. Glob Chang Biol 14:1531–1542

IPCC (Intergovernmental Panel on Climate Change).2000. Land use, land use change, and forestry: a special report of the IPCC. Cambridge, UK, Cambridge University Press.

Jose, S.2009. Agroforestry for ecosystem services and environmental benefits: an overview. Agrofor Syst.,76: 1–10.

Jung, M. 2005. The role of forestry projects in the clean development mechanism. Environmental Science & Policy,8: 87–104.

Lal R (2007) Carbon management in agricultural soils. Mitigation and Adaptation Strategies for Global Change 12: 303-322.

Lal, R., Follett, F., Stewart, B.A. & Kimble, J.M.2007. Soil carbon sequestration to mitigate climate change and advance food security. Soil Sci.,172: 943–956.

Lal, R., Kimble, J.M. & Stewart, B.A., eds. 2000. Global climate change and tropical ecosystems. Advances in Soil Science, pp. 341–364. Boca Raton, USA, CRC Press LLC.

Ramankutty, N., Evan, A.T., Monfreda, C. & Foley, J.A. 2008. Farming the planet: 1 Geographic dictribution of global agricultural lands in the year 2000.Global Biogeochem. Cycles, 22(1), GB1003.

Randerson, J.T., Chapin III, F.S., Harden, J.W., Neff, J.C. & Harmon, M.E.2002. Net ecosystem production: a comprehensive measure of net carbon accumulation by ecosystems. Ecol. Appl.,12:937–947.

Schlesinger, W.H. 1977. Carbon balance in terrestrial detritus. Ann. Rev. Ecol. Syst., 8: 51–81.

Shevliakova, E., Pacala, S.W., Malyshev, S., Hurtt, G.C., Milly, P.C.D., Caspersen, J.P., Sentman, L.T., Fisk, J.P., Wirth, C. & Crevoisier. C. 2009. Carbon cycling under 300 years of land use change: importance of the secondary vegetation sink. Global Biogeochem. Cycles, 23, GB2022.

Smith, P., Martino, D., Cai, Z., Gwary, D., Janzen, H., Kumar, P., McCarl, B., Ogle, S., O'Mara, F., Rice, C., Scholes, B., Sirotenko, O., Howden, M., McAllister, T., Pan, G., Romanenkov, V., Schneider, U., Towprayoon, S., Wattenbach, M. & Smith, J.2008. Greenhouse gas mitigation in agriculture. Philosophical Transactions of the Royal Society B - Biological Sciences, 363: 789–813.

Stevens, C.J. & Quinton, J.N. 2009. Diffuse pollution swapping in arable agricultural systems. Crit. Rev. Environ. Sci. Technol.,39: 478–520.

West, T.O. & Post, W.M. 2002. Soil organic carbon sequestration rates by tillage and crop rotation: a global data analysis. Soil Sci. Soc. Am. J.,66: 1930–1946.

Woodfine, A.2009. The potential of sustainable land management practices for climate change mitigation and adaptation in sub-Saharan Africa. Technical Report for TerrAfrica. Rome, FAO.

Zuidema PA, Baker PJ, Groenendijk P et al (2013) Tropical forests and global change:filling knowledge gaps. Trends Plant Sci 18:413–419

Chapter 7

CARBON SEQUESTRATION IN WETLANDS

S.S.Rakesh[1], R.Murugaragavan[2], R.Elangovan[3], and P.T.Ramesh[4]

1Doctoral Research Scholar in Environmental Science, Department of Environmental Sciences, Tamil Nadu Agricultural University, Coimbatore-641003

2Teaching Assistant in Environmental Science, Department of Soils and Environment, Tamil Nadu Agricultural University, Madurai-625104

3Assistant Professor in Soil Science and Agricultural Chemistry, College of Agricultural Technology, Tamil Nadu Agricultural University, Theni-625562, Tamil Nadu, India

4Associate Professor in Environmental Science, Agricultural College and Research Institute, Tamil Nadu Agricultural University, Killikulam-628252

INTRODUCTION

The Intergovernmental Panel on Climate Change (2013) estimates that cement production and fossil fuel combustion together have emitted 375 Pg of CO_2 in the atmosphere, whereas the change in land use pattern and deforestation released 180 Pg to the atmosphere. From 2006 to 2015, the average annual CO_2 emissions are estimated to be 10.3 ± 0.5 PgCyear^{-1} from various anthropogenic sources (Le Quere *et al.*, 2016). The atmospheric concentration of CO_2 was 390.5 ppm in 2011, and by 2015, it has increased to 400 ppm (WMO

2016). The longest record of continuous monitoring of CO_2 in the atmosphere at Mauna Loa, Hawaii, started in 1958 by C.D. Keeling and reported that the level has reached to 407 ppm by May 2018 (Dlugokencky and Tans, 2018).

Wetlands are described as "the lands of transition zone between aquatic and terrestrial ecosystems where the land is covered by shallow water" (Mitsch and Gosselink 1986). Wetlands represent only 3% of the world's soils, but account for approximately 21% of the global soil organic carbon (C) stock (Scharlemann *et al.,* 2014). Under natural, waterlogged conditions, slow decomposition favors the accumulation of soil organic matter, leading to a net C sink (Wilson *et al.,* 2018). Wetland acts as a sink for increasing the level of CO_2, thereby reducing carbon load and global warming (Whiting and Chanton 2001).

Wetlands are found in all climatic zones ranging from tropics to the tundra occupying about 5% of the earth's land area, wetlands are dynamic and natural ecosystems characterized by water logged or standing water conditions during at least part of the year. In most wetlands, water levels fluctuate seasonally instead of being stable, a property that accounts for making wetlands highly productive environments. Productivity among wetlands varies depending on the type of the wetland, climatic condition and vegetation communities. Along with productivity, decomposition is another complicated process that involves both aerobic and anaerobic processes.

Wetlands

The worldwide intergovernmental treaty on wetlands signed at Ramsar, Iran, in 1971 includes marsh, fen, bog, peatland or flowing water, static water, and fresh, salty, or brackish water whether artificial or natural areas of marine water (the depth should be maximum 6 m at low tide) into the wetland (Bowman 2002). However, in the Ramsar Convention, paddy fields, river channels, and anthropogenic water bodies are not comprised in wetland. As wetlands have zoological, ecological, botanical, hydrological, and limnological importance, they are categorized as "wetlands of international importance" underneath the Ramsar Convention (Stevenson and Frazier, 1999).

Post *et al.* (1982) reported that wetlands cover a total land area of 280 million ha worldwide, and the average carbon density in wetland is 723t per ha. This amounts to a total of 202.44 billion tons of carbon in wetlands of the world. Wetlands are one of the largest natural sources of the greenhouse gas methane (Bergamaschi *et al.*, 2007) yet at the same time, they have the best capacity of any ecosystem to sequester and retain carbon through permanent burial (Mitsch and Gosselink 2009). Carbon storage in wetlands has another implication. Of the total storage of organic carbon in the earth's soils, 20–30% or more may be stored in wetlands (Lal 2004) and is more vulnerable to loss back to the atmosphere as both carbon dioxide and methane if the climate warms or becomes drier.

Wetlands offer many ecosystem services to humankind, including water quality improvement, flood mitigation, coastal protection, and wildlife protection (Mitsch and Gosselink, 2009). It is estimated that 20–30 % of the Earth's soil pool of 2,500 Pg of carbon (Lal, 2009) is stored in wetlands (Bridgham *et al.*, 2006), although wetlands comprise only about 5–8 % of the terrestrial land surface (Mitsch and Gosselink, 2009). It has been estimated that different kinds of wetlands contain 350-535Gt C, corresponding to 20-25% of world's organic soil carbon (Gorham, 1998). However, the actual quantity of carbon stored in wetlands can only be estimated with a broad range of uncertainty. Hence, carbon fluxes and pools vary widely in different wetlands

Carbon sequestration in wetlands

Wetland ecosystems are characterized by the presence of stagnant water during least part of the year. This allows the development of specialized hydric roots and hydrophytic vegetation adapted to the presence of water and to the saturation of the soil (Reddy and De Laune 2008). They are known to provide an optimal natural environment for the sequestration and long-term storage of carbon dioxide from the atmosphere.

Carbon sequestration occurs through two main processes in fresh water wetlands which includes sediment deposition from uplands and on-site organic matter production, compared to the peatlands wherein carbon is sequestered only through on-site plant production (Bridgham *et al.,* 2006). Wetlands, no doubt, cover only 6–8% of the freshwater surface estimated to account for one-third of the world's organic carbon pool (Mitsch and Gosselink, 2009). Wetland ecosystems have unique characteristics as they are the sources of cultural, economic and biological diversity. These unique characteristics affect carbon dynamics and there are few mechanisms that aid in carbon storage in wetland ecosystem. In mechanism photosynthesis, wetland trees and other plants convert atmospheric carbon dioxide into biomass. Hence carbon may be temporarily stored in wetlands as trees and plants and the living material which feed upon them, and detritus including fallen plants and animals which feed upon them.

Tropical wetlands store 80% more carbon than temperate wetlands according to findings based on the studies conducted to compare ecosystems in Costa Rica and Ohio (Bernal and Mitsch, 2008). Tropical wetland in Costa Rica accumulated around 1 ton of carbon per acre (2.63t/ha) per year, while the temperate wetland in Ohio accumulated 0.6 tons of carbon per acre (1.4t/ha) per year (Bernal 2008). Many wetland plants are known to use atmospheric carbon dioxide for their main C source, and their death/decay and ultimate settlement at a wetland bottom can have profound effect on C sequestration. Even this mechanism of storage through photosynthesis depends along the latitudinal gradient as growth of vegetation is slow for high latitude wetlands with less sun, nutrient and colder temperature. Secondly, carbon rich sediment are trapped and stored that are brought along floods, hurricanes or even drained from watershed sources. However, long term storage is often limited due to rapid decomposition processes and rerelease of C to the atmosphere such as in case of paddy fields. Hence, wetlands are dynamic ecosystem where significant quantities of C from both wetland and non-wetland sources may also be trapped and stored in wetland sediments (Adhikari *et al.*, 2009)

The three main processes responsible for carbon sequestration in wetlands include photosynthesis or primary productivity, sedimentation, and nutrient enrichment through external factors (Miria and Khan 2014). Photosynthesis by producers is the main process which is responsible for the addition of all the organic matter to the wetland floor. Since, wetlands are highly productive ecosystems; their plants sequester carbon readily from the atmosphere and store it in their standing biomass. All of the organic carbons which find its way to the wetlands, either exogenously or endogenously, are manufactured by the plants (Kayranli *et al.*, 2010).

Exogenous sources include eroded soil material and terrestrial plant debris, whereas endogenous sources comprise of plankton and aquatic macrophytes. Nutrient enrichment from uplands in the form of eroded material which remains suspended in the water inflowing into the wetlands (Bridgham *et al.*, 2006). Carbon inputs mainly constitute carbon dissolved and suspended in inflowing waters and runoff (allochthonous) (Roner *et al.*, 2016), as well as carbon contained in organic matter from senesced vegetation in and around the wetland (autochthonous) (Alongi, 2014).

Forms of carbon sequestration in wetlands

Wetland carbon occurs in five main forms such as particulate organic carbon, dissolved organic carbon, plant biomass carbon, microbial biomass carbon, and gaseous end products such as methane and carbon dioxide. Except plant biomass carbon, all others are present in detritus, water, and soil (Kadlec and Knight, 1996). However, plant biomass carbon represents the active standing biomass, and it occurs in various life forms including emergent, submerged, or floating types.

Carbon cycle is comprised of many forms of soil carbon in case of wetlands such as plant biomass, standing dead plants, dissolved organic carbon, particulate organic carbon, and refractory carbon such as resistant carbon, which retains its strength even at high temperatures (Wynn and Liehr, 2001). Plant biomass is the active biomass, and it also includes periphyton (detritus and microorganisms attached to submerged surfaces). The particulate organic carbon comprises of particulate influent and organic substances, decaying plant material and microbial cells (Mustafa *et al.*, 2009). The dissolved organic carbon comprises of dissolved biochemical oxygen demand

and other carbon components in solution. Microbial biomass carbon occurs in heterotrophic microfloral catabolic activities, converting organic carbon back to its inorganic form and mineralizing dissolved organic carbon and particulate organic carbon. Gaseous forms of carbon are the result of either aerobic or anaerobic decomposition processes in the wetland soils.

Table 1: Current Carbon sequestration status in wetlands

S.No	Wetland type	g C m² year	Reference
1	General range for wetlands	20–140	Mitra *et al.* (2005)
2	Northern peatlands Peatlands (North America)	2	Gorham (1991)
3	Temperate peatlands	10–46	Turunen *et al.* (2002)
4	Boreal peatlands	15–26	
5	Californian Anderson tule marsh	106–155	Kim (2003)
7	Estonia(Europe) Freshwater marsh	15–2200	Mander *et al.* (2008)
8	Netherlands Peat meadow	280	Hendriks *et al.* (2007)
9	Brazil Brasileira	260	Bonotto and Vergotti (2015)
10	Denmark Reed marsh	504	Brix *et al.* (2001)
11	Temperate/Tropical Wetlands Coastal wetlands, North America		Chmura *et al.* (2003)
12	Mangroves	180	
13	Salt marshes	220	

14	Coastal wetlands, North America		Craft (2007); Craft *et al.* (2009)
15	Tidal freshwater wetlands	140 ± 20	
16	Brackish marshes	240 ± 30	
17	Salt marshes	190 ± 40	
18	Mangrove swamps, S.E. Asia	90–230	Suratman (2008)
19	Coastal wetlands, S.E. Australia		Howe *et al.* (2009)
20	Undisturbed sites	105–137	
21	Disturbed sites	64–89	
22	Tropical freshwater wetland	56 (for 24,000 years)	Page *et al.* (2004)
		94 (for last 500 years)	
23	Cyperus wetland in Uganda	480	Saunders *et al.* (2007)
24	China Changbai Mountain	200	Bao *et al.* (2010)
25	Prairie pothole wetlands, North America		Euliss *et al.* (2006)
26	Restored (semi-permanently flooded)	305	
	Reference wetlands	83	
27	Florida Everglades	86–387	Reddy *et al.* (1993)
28	Created temperate marshes, 10-years old, Ohio	181–193	Anderson and Mitsch (2006)
29	Tropical flow-through wetland, Costa Rica	306	Mitsch *et al.*, (2012)

Peatlands

Peatlands are globally significant terrestrial carbon (C) reservoirs and play a key role in global climate cycles (Yu, 2011). In peatlands, the imbalance between plant biomass production and decomposition results in the accumulation of thick C-rich deposits over millennia. Northern peatlands have sequestered an estimated 547 Gt C during the Holocene (Yu *et al.,* 2010), acted as persistent sinks of CO_2 and hence had a net cooling effect on the global climate historically (Frolking and Roulet, 2007). The C storage in the organic layers (peat) of the forested peatland sites documented here are much higher than those reported from temperate and boreal upland forests in eastern Canada estimated at 3.8 kg C m^{-2} (Marty *et al.,* 2015).

Peatland has been recognized worldwide as highly important for carbon storage since it accounts for nearly 50% of the terrestrial carbon storage with only 3% cover of world's land area (Guo and Li, 2007). Gorham (1991) calculated the pool in boreal and subarctic peatlands alone to be 460Gt. Whereas the carbon stored in peat could be 44-71% of the whole carbon held in the terrestrial biota (737 Gt), according to Matthews and Fung (1987).

Minerals associated with carbon sequestration

Reactive iron (Fe) and aluminum (Al) minerals are thought to contribute to soil C accumulation in soils through direct sorption of soil C to poorly crystalline minerals and/or the coprecipitation of soil in organo-metal complexes, particularly under aerobic conditions (Chen *et al.,* 2020; Kleber *et al.,* 2015). In contrast, Fe minerals can also contribute significantly to C loss via anaerobic respiration by Fe-reducing bacteria (Wagai and Mayer, 2007).

In upland tropical forest soils, microbial Fe reduction accounted for up to 44% of organic C oxidation from soils on an annual basis (Dubinsky *et al.*, 2010). As with Fe, Al can react with soil organic matter and facilitate C preservation (Porras *et al.*, 2017). The production of Fe(II) via anaerobic microbial respiration coupled to Fe-reduction is known to be a pathway of soil C loss in upland soils and fens (Bhattacharyya *et al.*, 2018; Emsens *et al.*, 2016). Knox *et al.* (2015) estimated net ecosystem greenhouse gas losses from the drained corn and pasture sites of 5.7 and 3.9 Mg C eq ha^{-1} year^{-1}. Heterotrophic respiration associated with Fe reduction thus accounted for 12% and 17% of these Ceq losses, respectively, assuming a 14-day redox cycle. For the reflooded wetlands, a 14-day Fe-redox cycle would oxidize 128% of the C emitted from methane fluxes (0.53 Mg C ha^{-1} year^{-1}) annually (Knox *et al.*, 2015). Previous studies have shown daily redox fluctuations in wetland soils. In upland soils, Fe(II)HCl may represent a proxy for the extent of soil anaerobic conditions (Hall and Silver, 2015).

Wetlands – Indian context

In India, wetlands cover about 4.7% of the total geographical area. These include about 18,154 natural and 9249 man-made wetlands representing about 5.31 m ha and 2.27 m ha of the total area, respectively (SAC 1998). Of this huge figure, only 26 wetlands have been designated as Ramsar sites with the world's total figure of about 2266 Ramsar sites (Finlayson *et al.* 2019).

Himalayan Wetlands

This type of wetlands includes the parts of Central Himalayas, Eastern Himalayas, Ladakh, and Zanskar (Pangong Tso, Chantau, Tso Moriri, Noorichan, Hanley, and Chushul marshes) and few portions of

Kashmir Valley (Dal, Anchar, Haigam, Kranchu, Malgam, Wular, and Haukersar lakes).

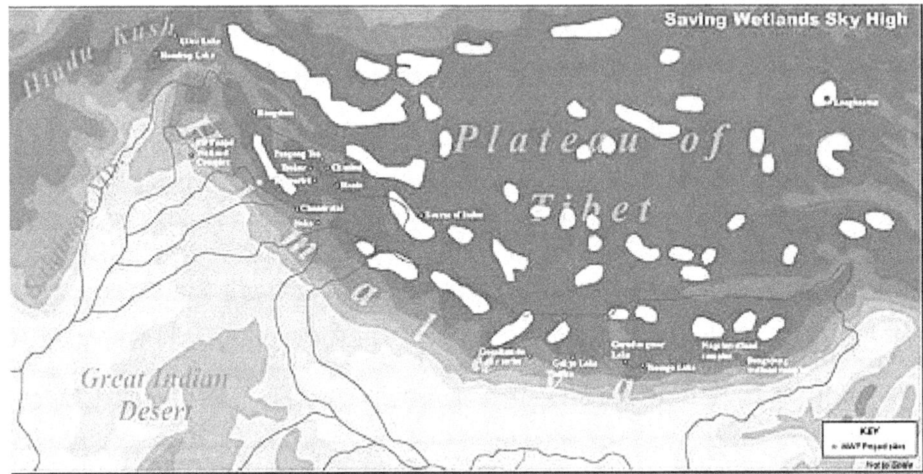

Source: WWF India

Indo-Gangetic Wetlands

Through the whole stretch from the river Indus at one end of west to Brahmaputra at the other end of east, there lies the largest wetland system in India called the IndoGangetic floodplain. The wetlands of the Indo-Gangetic plains and the Himalayan Terai are included in this type of wetlands.

Coastal Wetlands

The lagoons, mangroves, and massive intertidal expanses are included in the coastal type of wetlands. These are stretched along the 7500 km coastline in West Bengal, Orissa, Tamil Nadu, Andhra Pradesh, Karnataka, Goa, Kerala, Gujarat, and Maharashtra. This category of wetland includes Andaman and Nicobar Islands, Gulf of Mannar, Lakshadweep, Gulf of Kutch, and Sundarbans of West Bengal. 2.4

Deccan

Several tanks for storing water and numerous trivial and huge reservoirs along with few natural wetlands in nearly each town in the associated region are included in this category of wetland ecosystem.

Conservation

Conservation of the wetlands around the globe is very much essential in long term carbon storage and habitat for millions of birds, reptiles and mammals. Some of the factors responsible for the conservation of wetlands are as follows.

Fen depth

Fen depth has been necessary since we assumed that most of the organic material will soon be lost anyway at peat depths of less than 1 m.

Rewetting potential

The rewetting potential is chosen as a criterion because one has to be sure that sufficient water is available in the area to allow permanent flooding and the purpose of a wetland as a sink can be restored. Assessment of rewetting potential is specifically important and must include the entire catchment area of the wetland to be restored.

Management practices

Minimization of soil erosion is the primary reason of the most commonly applied practice, residue management, conservation crop management, no-tillage/strip tillage, conservation cover, afforestation, reduction of land use change, reduction of overgrazing, and increase in vegetative cover and irrigation (Faulkner *et al.*. 2011)

Presence of suitable target species

The third major criterion, the occurrence of target species, is more relevant when areas cannot be rewetted sufficiently. In that case, the existence of characteristic fen or fen meadow species is important for carrying out a more flexible plan, in which several development goals must be pursued simultaneously. It may take some time before the site conditions of the restoration area meet the Degradation and depletion of sea grass in a coastal wetland, which is often caused by erosion which leads to eutrophication or dredge-and-fill activities, is commonly restored by transplantation (Burkholder *et al.*, 2007). The suitable place and donor population for replacement should receive more focus on the significant expense (Bastyan and Cambridge 2008). Along with that, seeding techniques and mechanical planting have also been used as a possible solution to restore sea grass loss (Van Katwijk *et al.*, 2009).

To resolve the complication of degraded wetlands caused by *Spartina alterniflora* invasions in the Yangtze River Delta, some are the examples especially in Chongming Dongtan wetland in which methods that involve breaking of rhizomes, cutting, digging and tillage and waterlogging as well as biological substitution with *Phragmites australis* proved effective (Liu *et al.*, 2013)

Factors Affecting Carbon Sequestration in Wetland

- The availability of substrates and slow down OM decomposition and serve its accumulation in the soil include nutrient scarcity that limits the growth of microbes, high content of organic compounds with low degradability, and physical protection of organic particles through the formation of aggregates (Hernes *et al.*, 2020; Kuzyakov 2010; Six *et al.*, 2002).

- Temperature does not drive the decomposition of SOM as much as substrate availability does, as biological reactions are expected to double with every 10 °C rise in temperature (Hartel *et al.*, 2005). Northern wetlands are known to store over 50% of the global organic carbon due to slower rates of organic carbon decomposition because of cold temperatures and wet surface conditions (Hugelius *et al.*, 2013).

- Anaerobic conditions limit the enzymatic activity involved in SOM degradation, for two important reasons; firstly they need O_2 availability particularly in the case of phenol oxidases (enzymes capable of degrading recalcitrant materials such as humin and lignin). Secondly, the activity of SOM-degrading enzyme activity is inhibited by the compounds that get accumulated under anaerobic conditions.

- Large quantities of DOM can enter deep soil layers, getting retained in the wetland mineral soil and is, thus, sequestered efficiently. It mostly comes from root exudates or from the surface standing water and gets released into the soil pore water. The contribution of DOC depends upon its nature whether recalcitrant or degradable, and it is considered as the source of C for deep pools where plant roots mostly remain unreachable.

- Wetlands although act as buffers in hydrological cycle and as sinks for organic carbon, counteracting the effect of the increased CO_2 concentrations in the atmosphere. One of the interesting things about wetlands is their positive feedback related to climate change that could occur in near future if they are not managed properly. Climate change will affect wetlands in two fundamental ways: it will affect their functional capacity and will shift the geographical location of wetlands (Erwin 2009).

- Soil compaction would result in a direct loss of aggregation in surface soils (Warren, Nevill, Blackburn, & Garza, 1986), and directly inhibit O_2 diffusion through reduced soil pore space (Stepniewski, Gliński, & Ball, 1994). Soil C concentrations in the pasture sites are thus more likely to be influenced by an increase in anaerobic bulk soil conditions associated with compaction.

Conclusion

The wetlands and peatlands are one of the precious gift to mankind and acts as an ideal habitat for wide range of flora and fauna and offers diversified system in the globe. It was considered as a transition zone between terrestrial and aquatic ecosystem and it is our duty to conserve wetlands to conserve water resources, biodiversity, habitat protection and long term carbon sequestration.

References

Adhikari, S., Bajracharaya, R.M. and Sitaula, B.K., 2009. A review of carbon dynamics and sequestration in wetlands. *Journal of Wetlands Ecology*, pp.42-46.

Alongi, D.M., 2014. Carbon cycling and storage in mangrove forests. *Annual review of marine science*, 6, pp.195-219.

Anderson, C.J. and Mitsch, W.J., 2006. Sediment, carbon, and nutrient accumulation at two 10-year-old created riverine marshes. *Wetlands*, 26(3), pp.779-792.

Bao, K., Yu, X., Jia, L. and Wang, G., 2010. Recent carbon accumulation in Changbai Mountain peatlands, northeast China. *Mountain Research and Development*, 30(1), pp.33-41.

Bastyan, G.R., Cambridge, M.L., 2008. Transplantation as a method for restoring the seagrass Posidonia australis. Estuarine, Coastal and Shelf Science 79, 289e299

Bergamaschi, P., Frankenberg, C., Meirink, J.F., Krol, M., Dentener, F., Wagner, T., Platt, U., Kaplan, J.O., Körner, S., Heimann, M. and Dlugokencky, E.J., 2007. Satellite chartography of atmospheric methane from SCIAMACHY on board ENVISAT: 2. Evaluation based on inverse model simulations. *Journal of Geophysical Research: Atmospheres, 112*(D2).

Bernal, B. and Mitsch, W.J., 2008. A comparison of soil carbon pools and profiles in wetlands in Costa Rica and Ohio. *Ecological Engineering, 34*(4), pp.311-323.

Bhattacharyya, A., Campbell, A.N., Tfaily, M.M., Lin, Y., Kukkadapu, R.K., Silver, W.L., Nico, P.S. and Pett-Ridge, J., 2018. Redox fluctuations control the coupled cycling of iron and carbon in tropical forest soils. *Environmental science & technology, 52*(24), pp.14129-14139.

Bonotto, D.M. and Vergotti, M., 2015. 210Pb and compositional data of sediments from Rondonian lakes, Madeira River basin, Brazil. *Applied Radiation and Isotopes*, 99, pp.5-19.

Bowman, M., 2002. The Ramsar Convention on wetlands: has it made a difference. *Yearbook of International Co-operation on Environment and Development, 2003*, pp.61-68.

Bridgham, S.D., Megonigal, J.P., Keller, J.K., Bliss, N.B. and Trettin, C., 2006. The carbon balance of North American wetlands. *Wetlands, 26*(4), pp.889-916.

Brix, H., Sorrell, B.K. and Lorenzen, B., 2001. Are Phragmites-dominated wetlands a net source or net sink of greenhouse gases?. *Aquatic botany*, 69(2-4), pp.313-324.

Burkholder, J., Libra, B., Weyer, P., Heathcote, S., Kolpin, D., Thorne, P.S. and Wichman, M., 2007. Impacts of waste from concentrated animal feeding operations on water quality. *Environmental health perspectives*, 115(2), pp.308-312.

Chen, X., Zhu, H., Yan, B., Shutes, B., Xing, D., Banuelos, G., Cheng, R. and Wang, X., 2020. Greenhouse gas emissions and wastewater treatment performance by three plant species in subsurface flow constructed wetland mesocosms. *Chemosphere*, 239, p.124795.

Chmura, G.L., Anisfeld, S.C., Cahoon, D.R. and Lynch, J.C., 2003. Global carbon sequestration in tidal, saline wetland soils. *Global biogeochemical cycles*, 17(4).

Craft, C., 2007. Freshwater input structures soil properties, vertical accretion, and nutrient accumulation of Georgia and US tidal marshes. *Limnology and oceanography*, 52(3), pp.1220-1230.

Craft, C., Clough, J., Ehman, J., Joye, S., Park, R., Pennings, S., Guo, H. and Machmuller, M., 2009. Forecasting the effects of accelerated sea-level rise on tidal marsh ecosystem services. *Frontiers in Ecology and the Environment*, 7(2), pp.73-78.

Dlugokencky, E. and Tans, P., 2018. Trends in atmospheric carbon dioxide, National Oceanic & Atmospheric Administration, Earth System Research Laboratory (NOAA/ESRL).

Dubinsky, E.A., Silver, W.L. and Firestone, M.K., 2010. Tropical forest soil microbial communities couple iron and carbon biogeochemistry. *Ecology, 91*(9), pp.2604-2612.

Emsens, W.-J., Aggenbach, C. J. S., Schoutens, K., Smolders, A. J. P., Zak, D., & van Diggelen, R. (2016). Soil iron content as a predictor of carbon and nutrient mobilization in rewetted fens. PLoS One, 11(4), 1–17. https://doi.org/10.1371/journal.pone.

Erwin, K.L., 2009. Wetlands and global climate change: the role of wetland restoration in a changing world. *Wetlands Ecology and management, 17*(1), p.71.

Euliss Jr, N.H., Gleason, R.A., Olness, A., McDougal, R.L., Murkin, H.R., Robarts, R.D., Bourbonniere, R.A. and Warner, B.G., 2006. North American prairie wetlands are important nonforested land-based carbon storage sites. *Science of the Total Environment, 361*(1-3), pp.179-188.

Faulkner, S., Barrow Jr, W., Keeland, B., Walls, S. and Telesco, D., 2011. Effects of conservation practices on wetland ecosystem services in the Mississippi Alluvial Valley. *Ecological Applications, 21*(sp1), pp.S31-S48.

Finlayson, C.M., Davies, G.T., Moomaw, W.R., Chmura, G.L., Natali, S.M., Perry, J.E., Roulet, N. and Sutton-Grier, A.E., 2019. The second warning to humanity–providing a context for wetland management and policy. *Wetlands, 39*(1), pp.1-5.

Frolking, S. and Roulet, N.T., 2007. Holocene radiative forcing impact of northern peatland carbon accumulation and methane emissions. *Global Change Biology, 13*(5), pp.1079-1088.

Garg, J.K., Singh, T.S. and Murthy, T.V.R., 1998. Wetlands of India. *SAC, Indian Space Research Organisation, Ahmedabad.*

Gorham, E., 1991. Northern peatlands: role in the carbon cycle and probable responses to climatic warming. *Ecological applications, 1*(2), pp.182-195.

Gorham, E., 1998. Acid deposition and its ecological effects: a brief history of research. *Environmental Science & Policy, 1*(3), pp.153-166.

Guo, J. and Li, G.P., 2007. Climate change in Zoige Plateau marsh wetland and its impact on wetland degradation. *Plateau Meteorology, 26*(2), pp.422-428.

Hall, S. J., & Silver, W. L. (2015). Reducing conditions, reactive metals, and their interactions can explain spatial patterns of surface soil carbon in a humid tropical forest. Biogeochemistry, 125(2), 149–165. https:// doi.org/10.1007/s10533-015-0120-5

Hartel, M., Wente, M.N., Hinz, U., Kleeff, J., Wagner, M., Müller, M.W., Friess, H. and Büchler, M.W., 2005. Effect of antecolic reconstruction on delayed gastric emptying after the pylorus-preserving Whipple procedure. *Archives of surgery, 140*(11), pp.1094-1099.

Hendriks, D.M.D., Van Huissteden, J., Dolman, A.J. and Van der Molen, M.K., 2007. The full greenhouse gas balance of an abandoned peat meadow.

Hernes, P.J., Miller, R.L., Dyda, R.Y. and Bergamaschi, B.A., 2020. Vegetation vs. Anoxic Controls on Degradation of Plant Litter in a Restored Wetland. *Frontiers in Environmental Science*.

Howe, A.J., Rodríguez, J.F. and Saco, P.M., 2009. Surface evolution and carbon sequestration in disturbed and undisturbed wetland soils of the Hunter estuary, southeast Australia. *Estuarine, coastal and shelf science, 84*(1), pp.75-83.

Hugelius, G., Bockheim, J.G., Camill, P., Elberling, B., Grosse, G., Harden, J.W., Johnson, K., Jorgenson, T., Koven, C.D., Kuhry, P. and Michaelson, G., 2013. A new data set for estimating organic carbon storage to 3 m depth in soils of the northern circumpolar permafrost region. *Earth System Science Data (Online)*, *5*(2).

IPCC (2013), Climate change 2013: The physical science basis, Contribution of working group I to the fifth assessment report of the intergovernmental panel on climate change, Cambridge University Press, Cambridge

Kadlec, R.H. and Knight R.L. 1996. Treatment Wetlands. Boca Raton, FL: Lewis Publishers

Kayranli, B., Scholz, M., Mustafa, A. and Hedmark, Å., 2010. Carbon storage and fluxes within freshwater wetlands: a critical review. *Wetlands*, *30*(1), pp.111-124.

Kim, J.G., 2003. Response of sediment chemistry and accumulation rates to recent environmental changes in the Clear Lake watershed, California, USA. *Wetlands*, *23*(1), pp.95-103.

Kleber, M., Eusterhues, K., Keiluweit, M., Mikutta, C., Mikutta, R. and Nico, P.S., 2015. Mineral–organic associations: formation, properties, and relevance in soil environments. In *Advances in agronomy* (Vol. 130, pp. 1-140). Academic Press.

Knox, S. H., Sturtevant, C., Matthes, J. H., Koteen, L., Verfaillie, J., & Baldocchi, D. (2015). Agricultural peatland restoration: Effects of land-use change on greenhouse gas (CO_2 and CH_4) fluxes in the Sacramento-San Joaquin Delta. Global Change Biology, 21(2), 750– 765. https://doi.org/10.1111/gcb.12745

Kuzyakov, Y., 2010. Priming effects: interactions between living and dead organic matter. *Soil Biology and Biochemistry, 42*(9), pp.1363-1371.

Lal, R., 2004. Agricultural activities and the global carbon cycle. *Nutrient cycling in agroecosystems, 70*(2), pp.103-116.

Lal, R., 2009. Sequestering atmospheric carbon dioxide. *Critical Reviews in Plant Science, 28*(3), pp.90-96.

Le Quéré, C., Andrew, R., Canadell, J.G., Sitch, S., Korsbakken, J.I., Peters, G.P., Manning, A.C., Boden, T.A., Tans, P.P., Houghton, R.A. and Keeling, R.F., 2016. Global carbon budget 2016.

Liu, L., Zhao, X., Zhao, N., Shen, Z., Wang, M., Guo, Y. and Xu, Y., 2013. Effect of aeration modes and influent COD/N ratios on the nitrogen removal performance of vertical flow constructed wetland. *Ecological Engineering, 57*, pp.10-16.

Mander, Ü., Lõhmus, K., Teiter, S., Mauring, T., Nurk, K. and Augustin, J., 2008. Gaseous fluxes in the nitrogen and carbon budgets of subsurface flow constructed wetlands. *Science of the Total Environment, 404*(2-3), pp.343-353.

Matthews, E. and Fung, I., 1987. Methane emission from natural wetlands: Global distribution, area, and environmental characteristics of sources. *Global biogeochemical cycles, 1*(1), pp.61-86.

Mitra, S., Wassmann, R. and Vlek, P.L., 2005. An appraisal of global wetland area and its organic carbon stock. *Current Science, 88*(1), pp.25-35.

Mitsch, W.J. and Gosselink, J.G., 1986. Wetlands, 539 pp.

Mitsch, W.J., Gosselink, J.G., Zhang, L. and Anderson, C.J., 2009. *Wetland ecosystems*. John Wiley & Sons.

Mitsch, W.J., Zhang, L., Stefanik, K.C., Nahlik, A.M., Anderson, C.J., Bernal, B., Hernandez, M. and Song, K., 2012. Creating wetlands: primary succession, water quality changes, and self-design over 15 years. *Bioscience*, 62(3), pp.237-250.

Mustafa, A., Scholz, M., Harrington, R. and Carroll, P., 2009. Long-term performance of a representative integrated constructed wetland treating farmyard runoff. *Ecological Engineering*, 35(5), pp.779-790.

Page, S.E., Wűst, R.A.J., Weiss, D., Rieley, J.O., Shotyk, W. and Limin, S.H., 2004. A record of Late Pleistocene and Holocene carbon accumulation and climate change from an equatorial peat bog (Kalimantan, Indonesia): implications for past, present and future carbon dynamics. *Journal of Quaternary Science*, 19(7), pp.625-635.

Porras, R.C., Pries, C.E.H., McFarlane, K.J., Hanson, P.J. and Torn, M.S., 2017. Association with pedogenic iron and aluminum: effects on soil organic carbon storage and stability in four temperate forest soils. *Biogeochemistry*, 133(3), pp.333-345.

Post WM, Emanuel WR, Zinke PJ, Stangenberger AG (1982) Soil carbon pools and world life zones. Nature 298:156–159

Reddy, K.R. and DeLaune, R.D., 2008. *Biogeochemistry of wetlands: science and applications*. CRC press.

Reddy, K.R., DeBusk, W.F., DeLaune, R.D. and Koch, M.S., 1993. Long-term nutrient accumulation rates in the Everglades. *Soil Science Society of America Journal*, 57(4), pp.1147-1155.

Saunders, M.J., Jones, M.B. and Kansiime, F., 2007. Carbon and water cycles in tropical papyrus wetlands. *Wetlands Ecology and Management, 15*(6), pp.489-498.

Scharlemann, J.P., Tanner, E.V., Hiederer, R. and Kapos, V., 2014. Global soil carbon: understanding and managing the largest terrestrial carbon pool. *Carbon Management, 5*(1), pp.81-91.

Six, J., Conant, R.T., Paul, E.A. and Paustian, K., 2002. Stabilization mechanisms of soil organic matter: implications for C-saturation of soils. *Plant and soil, 241*(2), pp.155-176.

Stepniewski, W., Gliński, J. and Ball, B.C., 1994. Effects of compaction on soil aeration properties. In *Developments in agricultural engineering* (Vol. 11, pp. 167-189). Elsevier.

Stevenson, N. and Frazier, S., 1999. Review of wetland inventory information in Western Europe. *Global review of wetland resources and priorities for wetland inventory, Supervising Scientist Report, 144.*

Suratman, M.N., 2008. Carbon sequestration potential of mangroves in Southeast Asia. In *Managing forest ecosystems: The challenge of climate change* (pp. 297-315). Springer, Dordrecht.

Turunen, J., Tomppo, E., Tolonen, K. and Reinikainen, A., 2002. Estimating carbon accumulation rates of undrained mires in Finland–application to boreal and subarctic regions. *The Holocene, 12*(1), pp.69-80.

Van Katwijk, M.M., Bos, A.R., De Jonge, V.N., Hanssen, L.S.A.M., Hermus, D.C.R. and De Jong, D.J., 2009. Guidelines for seagrass restoration: importance of habitat selection and donor population, spreading of risks, and ecosystem engineering effects. *Marine pollution bulletin, 58*(2), pp.179-188.

Wagai, R. and Mayer, L.M., 2007. Sorptive stabilization of organic matter in soils by hydrous iron oxides. *Geochimica et Cosmochimica Acta, 71*(1), pp.25-35.

Warren, S.D., Nevill, M.B., Blackburn, W.H. and Garza, N.E., 1986. Soil response to trampling under intensive rotation grazing. *Soil Science Society of America Journal, 50*(5), pp.1336-1341.

Whiting, G.J. and Chanton, J.P., 2001. Greenhouse carbon balance of wetlands: methane emission versus carbon sequestration. *Tellus B, 53*(5), pp.521-528.

Wilson, B.J., Servais, S., Charles, S.P., Davis, S.E., Gaiser, E.E., Kominoski, J.S., Richards, J.H. and Troxler, T.G., 2018. Declines in plant productivity drive carbon loss from brackish coastal wetland mesocosms exposed to saltwater intrusion. *Estuaries and Coasts, 41*(8), pp.2147-2158.

Wmo, G. and Gwp, G., 2016. Handbook of Drought Indicators and Indices. *Geneva: World Meteorological Organization (WMO) and Global Water Partnership (GWP)*.

Wynn, T.M. and Liehr, S.K., 2001. Development of a constructed subsurface-flow wetland simulation model. *Ecological Engineering, 16*(4), pp.519-536.

Yu, Z., 2011. Holocene carbon flux histories of the world's peatlands: Global carbon-cycle implications. *The Holocene, 21*(5), pp.761-774.

Yu, Z., Loisel, J., Brosseau, D.P., Beilman, D.W. and Hunt, S.J., 2010. Global peatland dynamics since the Last Glacial Maximum. *Geophysical research letters, 37*(13).

Chapter 8

SEA LEVEL RISE AND ITS IMPACT STUDIES ON AGRICULTURE USING GEOSPATIAL TECHNOLOGIES

J. Ramachandran[1] M. Rajeswari[2] and K. Arunadevi[3]

[1]Teaching Assistant, [2]Professor and Head, Department of Agricultural Engineering, Agricultural College and Research Institute, Tamil Nadu Agricultural University, Madurai 624 105

[3]Assistant Professor (SWCE), Department of Soil and Water Conservation Engineering, AEC&RI, Kumulur 621712

INTRODUCTION

The climate change studies like global warming, rise in sea level, melting of glaciers, higher atmospheric CO_2 concentrations etc., are an enormous concern in all countries around the world which has a direct impact on the Agricultural crops from the stage of sowing, till the harvest. Geographic Information System (GIS), Remote Sensing and Global Positioning Systems (GPS) are emerging new, advanced and modern computerized geospatial technologies which help in the researchers to study about the changes in any environment without having any physical contact. This chapter presents how Remote Sensing and GIS helps in studying climate change and its impacts on agriculture.

SEA LEVEL RISE

The sea level rise is the best indicator of climate change than any other atmospheric variable. Agriculture being the backbone of India's economy, it is seriously affected by the rise in sea level in many parts of India. The sea level rise impacts can be noticed in Tamil Nadu and Andhra Pradesh along the Cauvery delta and Godavari delta, formed by the major rivers flowing in this area. The major problems like saltwater intrusion, soil erosion, and reduction in the availability of fishery resources are due to increased ocean temperature. Geographic Information System and Remote Sensing are the low cost technologies which help in the study of sea level rise, its causes, effects and its impacts on agriculture.

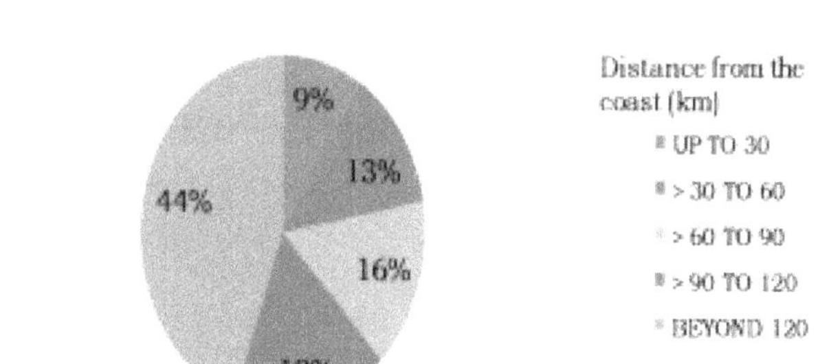

Fig. 1 Distribution of World Population as function of distance from nearest coastline

SEA LEVEL RISE AND AGRICULTURE

The past observations on the mean sea level along the Indian coast show a long-term (100 year) rising trend of about 1.0 mm/year. However, the recent data suggests a rising trend of 2.5 mm/year in sea level along Indian coastline. The sea surface temperature adjoining

India is likely to warm up by about 1.5–2.0° C by the middle of this century and by about 2.5–3.5° C by the end of the century. A 1 meter sea-level rise is projected to displace approximately 7.1 million people in India and about 5764 sq km of land area will be lost, along with 4200 km of roads *(Anupama Mahato, 2014)*.

Sea level rise predictions helps in the proper planning of agricultural crops. Sea level rise under different climate change scenario can be derived from the Model for the Assessment of Greenhouse-gas Induced Climate Change (MAGICC). MAGICC is a suite of coupled computer programs designed to assess the global-mean temperature and sea level changes that might arise from future emissions of greenhouse gases and of non-greenhouse gases that affect the lifetime of methane and of sulphur dioxide. Based on the output of MAGICC, the estimated sea level relative to the base year was derived for all years. The sea level rise estimate is associated with different scenario are computed and stored in Arc-Info.

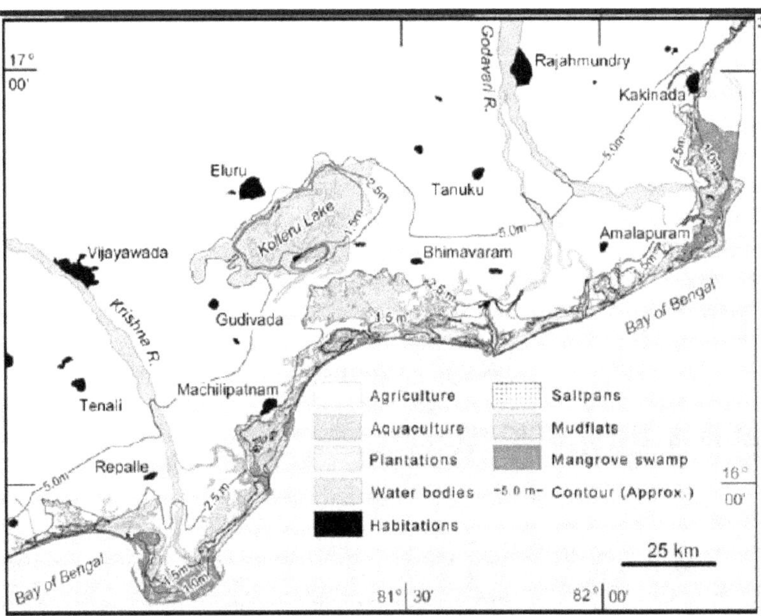

Fig. 2 Landuse Map of Andra Pradesh Coastal Regions

Sea level rise inversely affects agricultural activities in the coastal regions where the agriculture activities are intense. They used IRS-P6 LISS III data merged over SRTM elevation model. Based on the IPCC report the predicted Sea level rise, the high tide level and then by overlapping the predicted inundation (say 2.5m Sea Level) over the SRTM model, helps in determining the impacts like submergence of huge cropping land, plantation and aquaculture area over the coastal regions. The same was observed in deltaic regions of Krishna Godavari River at Andhra Pradesh *(Rao et al, 2009)*.

CONCLUSION

Studying the variation and prediction in the sea level in the field is a tedious and a time consuming process. But it is an important process that helps in saving the crop from the damages caused by the impacts due to sea level rise. Thus the Geographic Information System and Remote Sensing modules help in the prediction of sea level rise and crop planning accordingly.

REFERENCES

1. Anupama Mahato., (2014)., Climate Change and its Impact on Agriculture., International Journal of Scientific and Research Publications, Volume 4, Issue 4, ISSN 2250-3153

2. Rao, Nageswara, K., Subraelu P., Naga Kumar K.Ch.V., Demudu G., Hema Malini B., Ratheesh R., Bhattacharya S., Rajawat A.S. and Ajai., (2009). Geomatics Analysis of the Impact of Predicted Sea-Level Rise on the Agriculture along the Coastal Zone of Andhra Pradesh, Impact of Climate Change on Agriculture, 176-179